Solar Thermal Power Plants

Achievements and Lessons Learned
Exemplified by the
SSPS Project in Almeria/Spain

Written by F. G. Casal
Edited and with a Foreword
by P. Kesselring and C.-J. Winter

With 83 Figures

Springer-Verlag Berlin Heidelberg New York
London Paris Tokyo 1987

The Author

Dr. sc. tech. Federico G. Casal

Director, Rapperswil School of Engineering, Switzerland

The Editors

Dr. sc. nat. Paul Kesselring

Head, Prospective Studies Division, Swiss Federal Institute for Reactor Research (EIR),
Member of the SSPS Executive Committee,
Chairman of the SSPS Test and Operation Advisory Board

Professor Dr.-Ing. Carl-Jochen Winter

Member of the Board of Directors, Deutsche Forschungs- und Versuchsanstalt für Luft- und Raumfahrt e.V. (DFVLR),
Chairman of the SSPS Executive Committee

Library of Congress Cataloging-in-Publication Data
Casal, F. G. (Federico G.), 1925– Solar achievements and lessons learned exemplified by SSPS Project in Almeria/Spain.
1. Solar power plants—Spain—Almeria. 2. IEA-SSPS-Project. I. Kesselring, P. (Paul), 1932– II. Winter, C. J. (Carl-Jochen) III. Title
TK1056.C37 1987 621.31′244′094681 87–4879

ISBN 978-3-642-52283-3   ISBN 978-3-642-52281-9 (eBook)
DOI 10.1007/978-3-642-52281-9

Typesetting: F. Pustet, Regensburg

2362/3020-543210

Foreword by the Editors

The concept of constructing solar thermal power plants originated from the idea to replace the fossil fired by solar fired thermal power plants. Difficulties mostly encountered in such enterprises can be basically traced back to the fact that solar thermal power plants are more than just solar fired but otherwise conventional thermal power plants. Thus, when advancing from the present generation of experimental plants to the next generation of commercial operation demonstrating plants, more emphasis must be put on the solar specific aspects of non-solar specific subsystems and systems.

To address the question of solar energy future, three areas of relevant consideration will be treated

- the time constants of introducing new energy technologies,
- the primary energies in a post-fossil era,
- the aspects of future global trade and policy.

Since a typical time to introduce a new energy technology seems to be 30 to 50 years, continuity of progress is of highest importance. Therefore, a steady R+D program for solar energy is proposed emphasizing those sectors that promise high long-term impact. – A study of a 20–40 MW$_e$-Solar Thermal Power Station for erection and operation in a Southern European or Northern Africa country run by a group of European and American engineers is under way.

Where Are We Coming From?

In 1986, at the end of Phase I+II of the International Energy Agency's Small-Solar-Power-Systems-Project (The IEA-SSPS-Project) we are looking back at almost 10 years of worldwide R+D for solar thermal power generation. Other main activities besides SSPS are linked to names such as SOLAR ONE in the USA, EURELIOS (European Community) in Sicily, NIO in Japan, CESA I in Spain and THEMIS in France. What has been the motivation for such a R+D burst and what are the results of this effort till now?

When in 1973, as a consequence of the so called "first oil crisis", the idea of constructing solar thermal power plants originated, the concept was very simple and very convincing: Replace the fossil fired by the *solar fired thermal power plant!* The idea not only appealed to those efficienctly insolated countries that could use it in their own territory but also to some, which – from the very beginning – saw it mainly as a long term opportunity for their exports' industry. And, although the short term economic prospects were never over-whelmingly optimistic, economic competitiveness seemed not to be out of reach in the long run, considering increasing oil prices. – So much for the motivation.

On technical grounds, the concepts of the solar fired power plant led to a concentration of efforts to the solar specific subsystems, such as heliostat fields, receivers etc. The more conventional subsystems were thought to be available "from the shelf", i. e. to be taken over from conventional thermal power plant technology. As a result, the development of the solar specific subsystems which received a lot of attention turned out to be a technical success. Today, we have receivers with efficiencies around 90% and we have heliostat field availabilities and performances that are close to the design values.

The difficulties that were also encountered in these projects can basically be traced back to the fact that solar thermal power plants are *more than just solar fired but otherwise conventional* thermal power plants. The daily and annual solar cycles combined with the stochastic nature (clouds) of solar energy not only affect the solar specific subsystems but almost *any* subsystem of the plant. These problems can be handled, however, they have not been given enough attention during the past. Thus, when advancing from the present generation of experimental plants to the *next generation, demonstrating commercial operation,* more emphasis must be put on the solar specific aspects of non-solar specific subsystems and, above all, on the system as a whole.

Where Are We Bound to?

Oil is plentiful as well as cheap again and solar thermal plants – as solar energy in general – are not contributing a significant amount of energy to the world supply, as many too enthusiastic "solar freaks" had hoped 10 years ago. Should we forget about solar energy in general and solar thermal plants in particular?!

To address this question, we should touch at least three areas:

– time constants of introducing new energy technologies,
– primary energies in a post-fossil era,
– some aspects of global trade and policy.

Time Constants of Introducing New Energy Technologies

The basic ideas behind the presently used energy technologies are all quite simple:

- burn oil to produce heat,
- prepare an explosive gasoline air-mixture and ignite it in a controlled space in order to propel a piston,
- use some properties of certain unstable heavy nuclei (neutron induced fission, release of neutrons and energy in fission) to produce heat and consequently electricity in a controlled way.

Notwithstanding this basic simplicity, the development and market intro-duction of the technologies followed or still follow roughly logistic curves with characteristic times in the order of magnitude of 50 years – as can be read e. g. in relevant IIASA[1]-Papers. For comparison: The first chain reaction in a laboratory scale was in 1937; today, almost half a century later, the overall worldwide contribution of nuclear energy is 2,5% (1984) of the primary energy consumption; only in some heavily industrialized countries significantly more.

It is also worthwhile to remember, that most of this time is needed to *optimize the system* technically, economically and – at least for future systems – with respect to safety and environment. To build the first working burner, motor or experimental nuclear reactor was a quick shot compared to the rest that followed until a commercial product had a significant market share.

Assuming this point of view, it is not at all astonishing that solar energy – be it thermal or photovoltaic or other paths of conversion – has had no major impact on the world energy supply system during the first 10 years of development. Even if solar energy would turn out to be a brilliant success after an average introduction period of 50 years, this possibly would be difficult to recognize now: Only the most unspectacular *one fifth of the path is behind us, four fifth remain to go.*

Primary Energies in a Post-Fossil Era

Considering the long time constants just discussed, large scale solar energy utilization would become possible at a time, when fossil fluids might have become scarcer. On an even wider time scale, solar energy aside from coal and together with nuclear fission (breeder reactor) and nuclear fusion will be the only primary energies available, at least to the best of our present knowledge.

What could be a basis of comparison for these energy sources in order to assess the chances for solar energy? We choose the following three issues:

---

[1] IIASA – International Institute of Applied Systems Analysis, Laxenburg/Austria.

- material intensity, which is connected with problems such as energy-pay-back-time and cost,
- availability,
- functions and price of energy.

Material Intensity. Solar energy is often said to be too "dilute" and hence its use too material intensive in order to be a valuable technical large scale energy resource. In all probability this is not true and will be very improbable in the future. The reason ist that in principle not more than a few grams of aluminium per m² of reflecting surface are necessary to concentrate radiation to densities comparable to those in fossil flames or nuclear reactors (e. g. Megawatts/m²).

The decisive question is, of course, how much additional material has to be used in order to keep this surface properly oriented with respect to the sun and to protect it from wear and weather. Considering the presently discussed concepts of light weight designs, the plastic sheet heliostat shows that the trend actually goes away from today's heavy constructions. And furthermore, the installation cost share of the heliostat field decreased substantially over time: From 60% or more of the total investment of the plant in the early days of plants' development in the mid-seventies to almost 35% of modern plant lay-outs. That means that around two third of the plant's technologies do not belong to the so-called solar-specific items but to the more or less classic engineering field to be adapted to solar specifications. Certainly, there remains a large development potential in this respect.

Even for present day technology, however, the situation is not too bad. If an installed heliostat costs *200 $/m²* nowadays and collects only *1000 kWh/ m² of useful radiation per annum,* then it would have paid back the energy invested during its manufacturing in less than a year (in the form of concentrated radiation), provided less than *5 kWh/$* of heliostat cost[2] are involved in the manufacturing process. Of course, the energy-pay-back-time of the system as a whole will be longer because of the energy invested in the rest of the plant and the losses in transforming the concentrated radiation into the final product.

The above, by the way, shows also how important high efficiency applications of solar radiation are. We only have begun to learn the art of dealing with new type of high quality fuel with some unusual and fascinating properties. –

It is interesting to compare the energy-pay-back-times of different power plants: We know that energy-pay-back-times of coal plants are a few months and those of nuclear stations some months more; solar stations, for comparison, are estimated to have energy-pay-back-times of 2 to 3 years. – These figures are valid only, however, when counting the energy needed to

---

[2] This figure should not be mistaken as an – inverse – energy price or even the cost of radiation. The latter e. g. would be 2 to 4 ¢/kWh, assuming an annuity of 10 to 20%/a.

construct and erect the plants and to take them into operation. Not included are the energies related to the fuel cycle, such as prospection, mining, chemical treatment etc., or – concerning the spent fuel – treatment of exhausts, closing the nuclear fuel cycle or transportation of ashes, gypsum, and the like. This kind of additional energy consumption does not occur for solar power plants. Not to be forgotten in the energy accounting of any plant, however, is the decommissioning at the end of its lifetime.

In conclusion we may say that material intensity and energy-pay-back-times – although being important issues in further R+D definitely – present no obstacle to solar energy utilization.

Availability. Solar energy is renewable and abundantly available on a global scale. The problem is that energy is needed most, when and where the sun does not shine. Thus, *storage and transportation* is a decisive issue if it comes to the large scale use of solar energy.

Radiation is a kinetic form of energy. And, as the well known examples of a fly wheel and a hydro storage reservoir in the mountains show, kinetic energy is much more difficult to store than potential energy. Thus the transformation of the kinetic solar energy into a storable and transportable potential form becomes vital. As can be shown – and is made plausible by the almost ideal properties of oil in this respect – chemical energy seems to be most suitable for this purpose.

Heat and electric current are the two (kinetic!) forms of energy most often derived from solar radiation. The necessity to transform radiation, heat and electricity into chemical energy *(solar chemicals and fuels)* leads us to the conclusion that research in

– high temperature chemistry,
– electrochemistry and
– photochemistry

will be of utmost importance for the future. Only in such a transformed way, solar energy will get the high degree of availability – as well in time (storage) as in space (transport) –, necessary to take over some of the more valuable functions of oil on a global scale.

Functions and price of energy. Basically we are not paying for Kilowatthours (kWh) but for the function that a kWh fulfills, and we buy a given function as cheap as possible, of course (or alas?, thinking of how often this behaviour leads, e. g. to environmental problems). A good example are remote applications of photovoltaic cells, where the electricity produced costs often more than 1 $/kWh, but cannot be generated cheaper otherwise.

In a similar way it is clear, e. g., that traffic has also a very high functional value. The price of gasoline can be doubled several times before we will cease to drive cars and trucks completely. The interesting question is, which type of energy will be the cheapest substitute in this case.

Many people feel that solar energy has no chance at all compared to nuclear energy. This is – pricewise – certainly true at present. In the long run, however, the question seems more open to us – even if we generously assume that acceptance and additional safety measures, closure of the fuel cycle and waste disposal are not going to cost much more than anticipated now. This requires explanation. We use three arguments in favour of solar energy:

- Solar is a young technology with a considerable further cost reduction potential, as is shown, e. g., by the prospects of future heliostat cost development, given credit by the reduction correctly predicted and already achieved during the past ten years.
- Solar and nuclear technologies produce both mainly heat and electricity, i. e. kinetic forms of energy. Thus, if it is true that the conversion into storable and transportable forms of (potential) energy becomes important in the more distant future, this will apply to nuclear energy as well as to solar. It might be that then the cost of transformation, storage and transportation will be dominant as compared to the cost of the "raw energy". And this would be acceptable in a post fossil era if the functions provided by these forms of energy could not be bought cheaper. In such a situation the end energy prize would be influenced only marginally by the "raw energy" cost and other criteria, such as plant capacity (10 to 100 MW instead of GW blocks) or differences in acceptance, would gain importance.
- The local, dencentralized application of active and passive solar technologies leads to a rational energy use and – different from fossil or nuclear systems – to a tendency of minimizing the amount of energy supplied actively from the outside into the system. Thus a wider application of solar energy might have a strong influence on the energy demand spectrum and might reduce considerably the number and capacity of power plants required. Solar power plants will fit ideally into such a scenario.

As speculative as these considerations – concerning material intensity, availability and functions of solar energy in the future – may be, they show that it would not be wise to rule out "in a mood" solar energy as a candidate for future large scale use. Its potential is high, how uncertain its realisation may be for the time being. –

Some Aspects of Global Trade and Policy

Even more speculative than before, we sketch now a picture that might be a political motivation for large scale implementation of solar energy during the decades to come. The world is becoming interdependant more and more, inter alia by international trade. Balance of trade implies that partners have goods to sell and can afford to buy other products in return.

The world-wide north-south trade suffers already now from a heavy unbalance in favour of the north. The oil bill is one of the few positive items in the southern countries' bill. It will disappear in the post-fossil era. What can be done?

One of the few possibilities is to use the abundant sunshine in these countries to produce solar fuels and chemicals. These are export goods, with a high value added in the country and therefore provide the country with foreign currency. In contrast to other goods such as ores or other raw materials the "raw material sunshine" cannot be exported directly. This gives the guarantee, that the value added remains in the south, provided it gets the capital necessary to buy or produce the solar plants.

Solar energy thus inherently bears the possibility to contribute to a perpetuation of the world energy trade system which by and large served man sucessfully. A smooth bridging seems possible: Heat and electricity can be used onsite, electrolytically produced solar hydrogen might replace fossil fuels after their depletion.

Where, for example, is the difference between a substantial natural gas supply, say in the eighties of the 20th century in some cases transported over distances of several thousand kilometers through pipeline networks, paid for by annually contracted supplies of gas, and, sometime in the 21st century, a world solar hydrogen trade system using basically the same or similar pipeline transportation networks and the same basic contractual system between energy supplier country and energy consumer country?

Once More: Where Are We Bound to?

Having discussed the time constants, the possible role of solar radiation as an energy source in a post-fossil era, and some speculations on a future global solar energy trade, we come back to the original question. "Where are we bound to?", or more precisely "What do we need?".

A *parable* may paraphrase our present situation: In a forest there are the large old trees that are cut to provide us with the wood we need in our economy. There are also the medium aged trees that might be used in addition, if for some reason the supply of thick trunks would become insufficient. And last – but not least – there is undergrowth in every forest. Only a small number of its abundant population of very young trees will become the thick trunks, which we will rely upon decades from now. And it would be hopeless to try to identify them right now.

In this picture we compare fossil energies with the old thick trunks, which perhaps may not be used as extensively as before, e. g. for reasons of environmental impact. It is sometimes overlooked that 90% and more of worldwide used energy is fossil energy, fossil fuel of different kind, inevitably interfering with the environment when burned. – Nuclear energy is represented by the medium aged trees that can take over already if necessary and accepted. The energies called "alternative" and looked upon closer again during the past ten years, are the undergrowth of this "energy forest". Some of these technologies will be successful in the future, many of them will disappear. Only fools would derive from the forest analogy – i. e. the fact that only a very small percentage of young plants now available will become large trunks – the conclusion that the undergrowth should be eradicated instead of cultivated. –

It is also clear that it would end in a complete failure, if we would try to draw upon the undergrowth as a resource too early. Cultivating undergrowth is not an economically lucrative short term business, it is investing in the future. Thus, it becomes also evident that funds available cannot be abundant, but they should and may be sufficient without destroying the "forest economy".

The Importance of Long Term Aspects. Analogue statements can be made for the alternative energy sector. Obviously it would be foolish to expect a significant contribution from solar energy in general and solar thermal electricity generation in particular in the near future, i. e. by the year 2000 for example. On the contrary, raising too high expectations would lead to deceived hopes, and frustration would be the consequence. This type of vicious cycle has been experienced in the field of active solar heating and cooling technology during the last ten years. Initial overpromotion in the end led to an unjustified underestimation of this technology. – No need to repeat this bad experience.

However, it would be foolish also to deny the high potential of solar energy in the longterm and at least the possibility of its successful exploitation in the more distant future, merely by unduly overemphasizing the missing short term impact. As mentioned before, the typical introduction time for a new energy technology is 30 to 50 years. To repeat it once more, the most recent example is nuclear energy, which – starting from wartime R+D programs – covers now not more than a few percent of the world energy market. How could we expect more for solar energy?!

The Importance of Continuity. These considerations made, we propose a steady longterm R+D program for solar energy, emphasizing those sectors that promise high longterm impact. Funds may be moderate, e. g. an order of magnitude lower than for nuclear energy in the past. More important than the absolute amount of money available is the prospect that *good* projects will not be cut, e. g. because solar energy becomes less fashionable temporarily. Short term, high amplitude oscillations of enthusiasm, political support and funds – as experienced worldwide during the past 10 years – are not creating a climate, which is attractive to good scientists and engineers. In order to promote high quality, cost effective research, we do not need crash programs. What we need is continuity!

Elements and Realisation of a Long Term Solar Thermal Program

If it is accepted that continuous long term solar energy R+D is an issue, then the question arises which elements should a corresponding program contain and what the measures for its realization should be. We try to scetch answers on four different levels, namely

- R+D topics,
- organization,
- Solar Thermal Demonstrator Power Station,
- politics.

R+D topics. During the past ten years we have acquired a good understanding concerning topics such as

- low temperature and medium temperature systems producing process heat,
- importance of site characteristics such as meteorology, available infrastructure etc.,
- solar specific subsystems such as heliostat – or collector fields, receivers of electricity generating systems etc.

In particular we have learned that highly concentrated solar radiation from heliostat field is a "high quality fuel" which is already now not too far away from being commercially competitive. This leads to three important future R+D areas:

- Reducing further the amount of material used and the cost of solar concentrators in order to reduce the "fuel cost" further.
- Optimization of existing systems (such as solar thermal power plants) in order to make best use of the solar fuel. Here remains much to be done, mainly by the companies already involved in the field.
- Finding new applications which transform solar radiation into a storable and transportable form.

The last item seems of particular importance to us, for reasons discussed already earlier. *Solar Fuels and Chemicals is the keyword.* We remember that the R+D fields related to it are

- high temperature chemistry,
- electrochemistry and
- photochemistry.

Organization. The R+D sketched above takes place under precommercial conditions, it is "cultivating the undergrowth of the forest". As has been said before, resources are limited for this purpose. The advantage is that in this phase of the development there is not much competition, since there is not yet a real market. This means that potential future competitors all sit together in the same boat – for the time being. Their common objective is to learn as much as possible about new technologies – which might become competitive in the future – for a restricted amount of money.

This scenario is the ideal context for *international cooperation,* e. g. within the IEA, as has been the case with the SSPS-project. The 200 Mio DM invested in the "Plataforma Solar de Almeria" (IEA-SSPS, CESA I)[3], (GAST)[4] have created the core of what might become the International Solar Laboratory of Southern Europe. Large scale experiments – indispensable for the technologies considered – may be prepared by theoretical work and laboratory experiments on a national or multinational scale in the different countries. The cheapest place to realize them could be – at least for European countries – a well organized Plataform Solar in Almeria.

Three points are essential for the success of such a laboratory:

– There must be a sufficient number of qualitatively good experiments and
– we refer to what we have said under the Importance of Continuity in R+D.
– There must be a small, highly qualified and well equipped permanent staff on site. Its task is very challenging: It has to
  ○ prepare the installation of experiments,
  ○ install and conduct experiments,
  ○ collect data and evaluate experiments in cooperation with the experimentors.

Two ways seem feasible to fund such a facility:

– Base load funding by an international community of sponsors covering e. g. the salaries of the staff and the maintenance of the infrastructure. Cost related to a specific experiment would be carried by that experiment.
– The facility could be managed as a "profit center", i. e. the full cost would have to be born by the experimentors. National subsidies would be allocated rather to the experimental program than to the facility.

Solar Thermal Demonstrator Power Station

As it was usually the case with new technologies in the area of conventional energy conversion systems, viz. to erect medium scale demonstration facilities under realistic technical and geographical conditions (nuclear stations in the 200 MW range, coal gasification or coal liquifaction plants etc.), time seems to have come to step over from solar thermal laboratory or small scale development and test facilities to a medium scale *"Solar Thermal Demonstration Plant":* An international venture is suggested to put all lessons together, learned so far in the worldwide six major experimental

---

[3] CESA – Central Electro Solar de Almeria
[4] GAST – Gasgekühltes Solarturmkraftwerk

facilities, and erect a first Solar Thermal Demonstrator Power Station in the capacity range of $\leq 100\,MW_{el}$. A highly insolated energy supplying country shall be involved. –

Political Aspects. Worldwide, energy related problems are hotly debated issues in almost any political gremium. Life would be easier, if we could foresee our energy future with at least some certainty. Recognizing the fact that within one single decade the area of highest concern shifted from energy supply to such a different topics as environmental impact problems should remind us, how far we are away from having a "crystal ball" revealing "the truth". This is the basic reason, why we plead for a redundant energy R+D approach. Cultivating the "undergrowth" of the energy system gives the flexibility, necessary to accomodate to unpredictably changing boundary conditions.

Following different side lines to the main stream of the energy system development leads to a collection of fall back options, one of which may become a main path in the future. This procedure not only bridges the long time constants inherent to energy R+D, it is also cost effective because the unavoidable mistakes are made on a small scale. This contrasts favourably to the elimination of finally unsuccessful options in a crash program, necessary if the main route has to be changed in the absence of well investigated fall back options.

We strongly feel that Solar Thermal High Temperature R+D is one of the more promising side lines that should be followed carefully and continuously. In order to do that, we need political support for a Long Term Solar Program as sketched above.

Summary and Conclusions

We have shown that a worldwide 10 years R+D effort for solar thermal power plants has been a full technical success with respect to the solar specific subsystems. More emphasis must be put in the future on the optimization of the power plant system as a whole.

Companies involved in the past development feel ready to go for a commercial size demonstrations plant. Such a plant should be realized in the near future, if necessary for funding reasons, by an international joint venture group.

Concentrated solar radiation is a high quality fuel that may be used for many other high temperature process heat applications than solar electricity generation. The time has come to put more emphasis on R+D in these new fields. A particularly important long term issue is the conversion of solar energy into a storable and transportable form (solar fuels and chemicals).

The Plataforma Solar in Almeria – site of the SSPS' and CESA I plants – has the potential to become the solar thermal Laboratory in Europe for advanced large scale experiments in solar high temperature R+D. The viability of this option must be checked internationally and – if possible positive – decisions taken soon. Considering the worldwide limited funds, international cooperation – e. g. in the framework of IEA – is indicated.

Solar energy, nuclear fission and fusion are the prime energy sources of a postfossil era. Time constants in energy R+D are as long as 50 years. Therefore politicians should recognize the value of a – moderately funded – continuous long term solar R+D program. In order to promote high quality, cost effective research, we do not need crash programs. What we need is continuity.

About this Book

Much of what has been said in the above introduction is related intimately with our experience made in the Small Solar Power Systems (SSPS) Project of the International Energy Agency (IEA). This experience is documented in a complete "information pyramid". Its base is formed by technical reports (not available to the general public) followed by 4 volumes published by Springer (Final Report of the "International Test and Evaluation Team" including a "Book of Summaries", Springer, 1986) and the Final Report of the Operating Agent of the Project (SSPS SR-7, 1985). These documents are listed at the end of this book.

All these reports were written by "insiders" of the SSPS Project. Therefore, the SSPS Executive Committee found it appropriate to hire an external expert and ask him – as an "outsider" – to do two things:

– condense the nearly 2000 pages of available information to a size and form which makes it readable and valuable for an advanced student or the average technical engineer;
– "step back" and try to make a few "snapshots", illustrating the present status and the future possibilities of Solar Thermal Power Station Technology; in general, as exemplified by the lessons learned in the SSPS and other experimental projects.

The present book is the result of this effort and, therefore, represents the "top of the information pyramid". On behalf of the Executive Committee, the editors have followed and guided the work of the author, Prof. F. G. Casal. We thank him for his competent and efficient way of fulfilling the demanding task.

P. Kesselring · C.-J. Winter

Preface by the Author

In nature as well as in technology, energy is the key to life and therefore "conditio sine qua non" to our survival. Early man roamed fields and forests in mild climates, surviving by accepting goods and sources of energy which were offered by an abundant nature; today's sophisticated inhabitants of our planet depend for their survival in every source of energy which human ingenuity can discover. The consequence of this development has been an almost reckless use of non renewable resources and an equally thoughtless accumulation of waste materials; long range planning efforts must therefore seriously consider the possibility of substituting non renewable sources of energy with more renewable ones.

The amount of solar energy delivered to our planet every year is more than ten thousand times greater than the amount used for technical purposes during the same period of time. A large part of this energy is needed to run the global weather machine and a smaller part is used to sustain the natural vegetation and our agriculture. A considerable portion of this energy resource could be used for the production of electricity or as process heat without endangering the global environment. The optimistic solar power proponent will point to these facts and will state that the remaining technical difficulties will be solved with sufficient research and development funds. He must, however, also face up to non technical obstacles:

- fossil and nuclear resources will be sufficient to cover the energy needs of our technological culture for many decades to come,
- the use of fossil and nuclear resources, and of hydroelectric power for the purpose of generating useful energy is economical for the richest and for the poorest countries,
- the threat of the greenhouse effect due to the increase of atmospheric pollution with carbon dioxide, methane and other waste products is based on laboratory experiments and theoretical models alone. An absolute proof does not yet exist.

The essence of the long range planner's answer will most likely be:

- the natural renewal of mineral resources is far too slow to compensate for the rate at which we use them,
- the economics of mineral coal, oil and natural gas are subject to local and global political processes which are most likely outside the user's control,

– reliable proof of the greenhouse effect can only be furnished by letting it
happen, thus risking an uncontrollable situation which could easily turn
out to be the demise of human culture as we know it today.

However, the major obstacle of all long range efforts is the "we don't need it
now" syndrome. Every intelligent person will agree that what we do not
need now may be needed some time later, but many intelligent people forget
that it took roughly half a century before the elementary knowledge of how
to use coal, or oil, or uranium was developed to the point which allowed the
full-scale use of these resources for technological purposes.

The beginning of this century brought exhaustive knowledge on the use
of fossil energy sources; the middle of the century presented the same for
nuclear power sources. In terms of solar power plants this means that there
will be a time lag of decades before the time of sound knowledge gives rise to
a wide application of solar power technology. It is this time lag which makes
it unwise to make development decisions based on present energy market
economics or on present politics. Stable oil prices in the eighties are no
assurance for stable oil prices in the coming decades.

Another obstacle is the commendable intention to prevent the waste of
development funds. It is precisely here that solid and extensive analyses are
most important. Seemingly self evident facts and premature judgements are
a poor basis for separating promising from wasteful research projects. The
history of science and technology gives ample proof of the difficulties of such
judgements and even Nobel prize winners are not exempt from this pitfall:
Rutherford considered it impossible to exploit the binding energy of the
atomic nucleus for practical purposes and qualified as "fools" those who
attempted to do what Fermi proved feasible just a few decades later.

Any long range planner runs the risk of finding himself one day in a dead
end street. It cannot be over emphasized, however, that he is under a
constant obligation to consider ALL the facts – in the case of solar energy
not only the basic physics of the generation of power – and whenever
necessary to supplement his analyses with experimental or operational
findings. This is particularly true in cases where large quantitative uncertainties are attached to parameters of either technical or economic nature.

The purpose of this book is to shed light on such uncertainties and to
point out ways of reducing them.

F. G. Casal

Acknowledgements

This book is a condensation of a number of papers and publications written in connection with the experiences gained in the construction and operation of solar thermal power plants, with particular attention to one such project. The Executive Committee of the Small Solar Power Systems Project ("SSPS") of the International Energy Agency ("IEA") decided to make the experiences gained with this project accessible to all interested parties. It selected an outsider as author and put him under contract with the German Aerospace Research Establishment ("DFVLR") as the Operating Agent. After four hectic months I wish to express my gratitude to the Executive Committee for providing me with such an interesting experience.

I am especially indebted to Mr. Wilfried Grasse, project director at SSPS and my contact person at DFVLR in Cologne (Germany); he and his co-workers were instrumental in supplying the necessary documentation and in supporting my efforts in many ways. Dr. Paul Kesselring of EIR provided close support and also established necessary contacts with many of the key persons involved in the development of solar power plants, and his efforts made it easy to gain the broad perspective necessary to write such a book. Prof. Dr. C. J. Winter (DFVLR Stuttgart) contributed significantly with an exchange of views and ideas about the possibilities and the desirability of future research in solar utilization technology.

I am also indebted to Mr. Luis Crespo for giving me an overview of all development efforts going on at the "Plataforma Solar" in Almeria, Spain. Mr. Ricardo Carmona, member of the International Test and Evaluation Team ("ITET"), Mr. Pierre Wattiez, Mr. Hans Fricker and Mr. Mats Andersson deserve special thanks for having been able to supply information on short notice. I also want to thank all other individuals involved in the SSPS project who supplied necessary and more detailed information as well as the people at the GEORGIA POWER Company, at the SANDIA Corporation, at ARCO SOLAR, at SOLAR ONE and at the SOLAR DIVISION of the McDonnell Douglas Company, for their helpful and highly useful discussions.

Dr. Jerome Weingart must be included among those who contributed to the discussions concerning the future potential of the utilization of solar power and the recommendations for further work; he also supplied the glossary. My thanks are also in order for Wendy Tamborero, whose assistance in applying the finishing touches is highly appreciated.

Finally, I wish to express my gratitude to the Board of Trustees of the Rapperswil School of Engineering (Switzerland) for allowing me to go on leave for the necessary period of time.

Contents

1 Introduction

1.1 Historical Background and Relationship to the IEA

One of the objectives of the energy research, development and demonstration program of the International Energy Agency (IEA) is to promote the development and application of new and improved energy technologies which could potentially help cover our energy needs. Early in 1976, a working party for Small Solar Power Systems (SSPS) was created with the approval and encouragement of the Committee for Research and Development of the International Energy Agency (IEA) [1]. At that time the following countries showed interest in attending the formative meeting: Austria, Belgium, Canada, Great Britain, Greece, The Federal Republic of Germany, Italy, Japan, Spain, Sweden, Switzerland and the United States of America.

In its first meetings the SSPS Working Party explored the technological possibilities of the exploitation of solar power at small levels (photovoltaics, wind, waves and thermal power conversion) and also reviewed what was being done at that time in the domain of solar power in each of the participating countries. At a meeting in mid 1976 in Vienna, a study performed by MBB was presented. It stated that as distributed systems (systems using a large number of parabolic trough collectors "DCS", see chapter 4) grow linearly in terms of power, the associated costs grow as a function of the size of the intended system. By comparison, the cost per unit output of the central receiver systems ("CRS", see section 5.2) is expected to level out for very large sizes. The conclusion was that the cost per kW of installed electrical output of the two systems was expected to have a crossover near 500 kWe.

Since none of the solar components had yet been manufactured and since none of the involved companies had solid experience in solar technology, all cost numbers were gross estimates not based on solid knowledge. In order to evaluate the anticipated crossover of costs, a project to build both systems at the power level of 500 kWe was proposed to the working party, letting the member nations share the financing. Early in 1977 the discussions centered on the possibility of building only one or both systems, on the choice of potential sites as well as on the possible choice of the Federal Republic of Germany as the lead country for the project. Based on site possibilities and on the promise of administrative support, Spain was selected to be the host country for the project. In 1977 many engineers from 7 interested countries participated in a workshop to identify suitable working fluids for the CRS. Among the three competing working fluids (sodium, molten salts and air or gas), sodium was finally selected for thermal transfer and storage; however, this choice was not unanimous.

The Implementing Agreement was signed in the Paris headquarters of the IEA in October 1977, officially starting the project. Member countries were then Austria, Belgium, The United Kingdom, Greece, Federal Republic of Germany, Italy, Spain, Sweden, Switzerland and The United States. An Executive Committee (EC) was defined and the DFVLR (German Aerospace Research Establishment, see glossary) was appointed as the operating agent.

1.2 Project Implementation Within the IEA

In order to enable the project management to operate with a high degree of autonomy [SR 7, p. 20], the project organization shown in Fig. 1 was adopted.

The DFVLR thus served as the Operating Agent responsible for carrying out the SSPS project on behalf of the participating countries [IV,1.] and was also charged with the directorate of the plant.

In 1980 the DFVLR contracted the regional Spanish utility Compañia Sevillana de Electricidad (CSE) as the Plant Operating Authority.

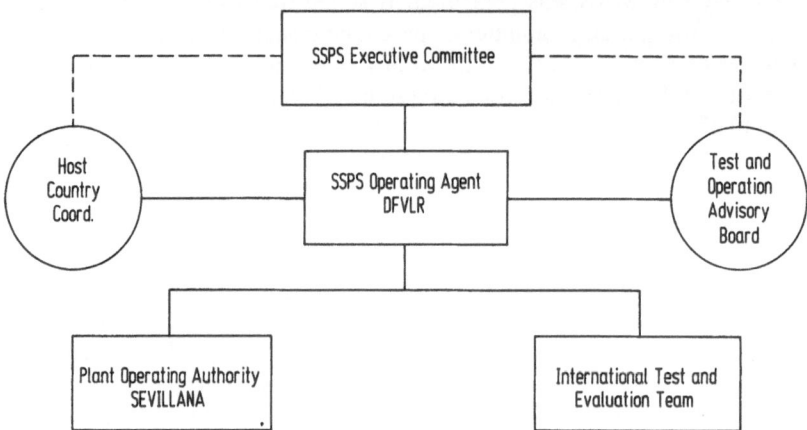

Fig. 1. Project organization of SSPS within IEA

In 1981 an International Test and Evaluation Team (ITET) composed of experts from the participating countries was made responsible for the scientific testing and for the evaluation of the work. The ITET was established by the Executive Committee which designated its head. Working on site, the ITET evaluated and reported on test and operation activities and recommended and advised the plant director on defining, planning, preparing and conducting tests and operations. The evaluation consisted in comparing and combining measured, calculated and reported plant data to determine the plant's performance and behavior over the entire period of the program.

A request for proposals for both the CRS and the DCS systems was sent out to industrial companies in all the interested countries. Out of the 12 industry proposals received and evaluated in 1978, the choice fell on INTERATOM / Martin Marietta / CASA for the CRS and on ACUREX / MAN / Tecnicas Reunidas for the DCS to design the proposed systems in detail and to perform a cost analysis. The design specifications, the actual design and the determination of the costs of construction were accomplished during 1978 under the leadership of the DFVLR as the Operating Agent and with strong support by a staff from the participating countries.

1.3 Specific Objectives

The main goal of the SSPS project [IV,1.] was to investigate two different possibilities of producing electricity at small power levels. In order to achieve this goal, two dissimilar types of solar thermal power plants were designed, constructed, tested and operated: a distributed collector system (DCS) and a central receiver system (CRS). They were built adjacent to each other on the "Plataforma Solar" (Spanish for "solar test site") in the province of Almería, in south eastern Spain both power plants have the same electrical output (500 kW design rating at equinox noon) and have delivered electric energy to the Spanish grid during the years 1981 to 1985.

In addition to examining in detail the feasibility of using solar radiation to generate electrical power, the project also included the following objectives [IV,1.]:

- to promote cooperation between IEA members in the development of new technologies,
- to demonstrate the technical feasibility of building solar power plants with available hardware,
- to design a plant which operated at 500 kWe, but with the potential of being scaled up or down,
- to assess investment costs while achieving reasonable operating expenses, good engineering safety and a long lifetime,
- to gather operational performance data of such power plants,

– to evaluate the viability of the DCS and the CRS concepts,
– to assess the potential of further solar power plant development.

In the end, the project was conducted under the auspices of the IEA by nine of its member countries: Austria, Belgium, Switzerland, Germany, Spain, Greece, Italy, Sweden and The United States of America; Great Britain having elected not to participate in the construction and test phases of the project.

In spite of all setbacks and disappointments inherent to projects of this nature, the majority of the objectives was attained; in particular, the solar specific subsystems have performed close to expectations.

2 Description of the SSPS Site

2.1 Criteria Leading to Its Choice

A site near the village of Tabernas (in the Spanish province of Almería) was chosen [III,2.] for the construction of the two power plants. This site is also utilized by Spanish organizations interested in the development of solar technology and has therefore been given the name "Plataforma Solar". The geographic location of the SSPS plants is 2° 23'W and 37° 06'N and the elevation is 500 m above sea level (Fig. 2).

Factors influencing this choice were mainly the following:

- good solar statistics: 2950 h/a sunshine at the Almería airport. The yearly global energy received on a horizontal surface was estimated to be between 1700 and 1800 kWh/m²,
- good transportation to the site: the city of Almeria is a international port

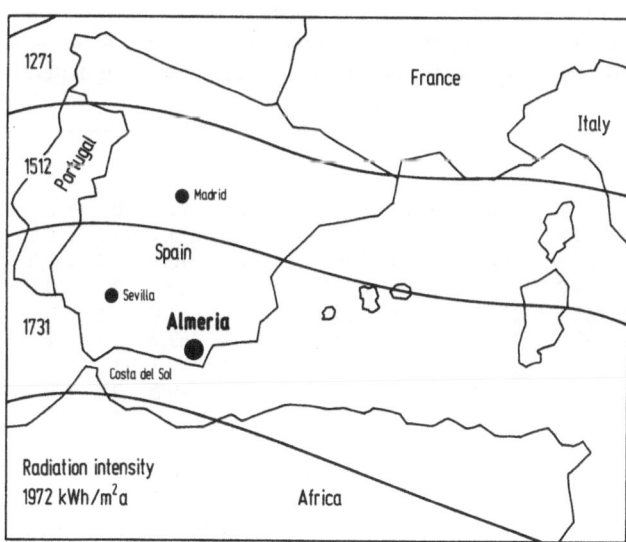

Fig. 2. Location of SSPS site in Spain

and has an international airport capable of receiving cargo aircraft. The distance from the site to this airport is roughly 40 kilometers,
- poor agricultural value of the land: frequent efforts to improve this semiarid area have had no significant success.

2.2 Characteristics of the Site

In order to assess the adequacy of the site for the technical utilization of solar energy, the conventional meteorological data have to be amplified with measurements of specific parameters which are very important for this particular application. These parameters are mainly:
- the insolation as the source of energy with particular attention paid to its statistical distribution during a typical year,
- wind as a force perturbing the alignment of the heliostats,
- soiling and cleansing factors affecting the reflectivity of the mirrors.

2.2.1 Insolation

Solar radiation data were first obtained on the site during the period from April 1978 to April 1979 [TR 1/81] and these data were supplemented by the insolation data gathered during construction and operation of the two plants [III,3.1]. Since the site is some 40 km to the north of the Almería airport and is separated from it by a range of mountains, the actual insolation data were found to differ significantly from the simple solar statistics observed in earlier years at the airport. In particular, it is important to take into account that global radiation is not a measure of adequcay for solar thermal power plants, since the diffuse component of the radiation does not contribute to the yield of focusing devices; only the radiation on a surface perpendicular to the beam direction ("beam radiation") is useful.

W/m^2 month	>300	>400	>500	>600	>700	>800	>900	>920
Jan	204.3	186.7	168.5	138.3	102.5	32.5	0	0
Feb	147.4	134.3	118.1	99.6	70.5	21.3	0	0
Mar	240.9	219.8	192.3	161.9	118.2	73.4	14.3	2.9
Apr	239.7	214.6	190.4	161.1	124.8	76.2	8.4	3.2
May	290.7	270.8	244.4	212.7	171.2	108.3	7.3	0.5
Jun	261.1	230.4	189.8	138.2	84.0	37.0	1.6	0
Jul	339.5	314.3	278.6	228.9	153.9	69.4	7.5	2.2
Aug	297.8	274.0	248.8	213.6	155.0	89.7	11.5	3.7
Sep	278.7	260.2	186.3	186.3	131.4	69.2	9.4	0
Oct	205.7	189.9	172.2	137.4	93.1	44.9	11.6	5.7
Nov	127.9	116.7	101.8	82.4	60.0	27.8	1.6	0
Dec	187.2	172.6	154.4	133.4	106.8	62.1	7.0	0
total	2820.9	2584.3	2288.5	1893.8	1371.4	711.8	80.2	18.2

Fig. 3. Monthly hours of beam irradiance during 1983 [III,3.1]

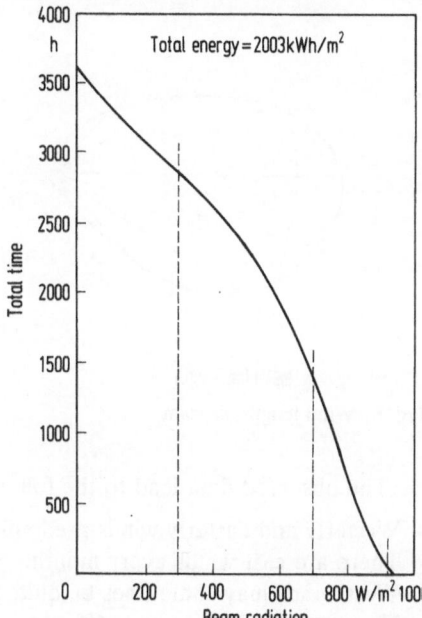

Fig. 4. Monthly averages of beam irradiation

The sources mentioned above were used to obtain an estimate of the variance of the solar data obtained from the Data Acquisition Systems ("DAS") by the evaluation team; the values presented in Fig. 3 and used to obtain the graph shown in Fig. 4 were selected as typical.

As Figs. 3 and 4 show, the site has excellent beam irradiance statistics for $300\,W/m^2$. The statistics around $700\,W/m^2$ are good with some 1300–1400 hours of beam radiation supplied per year.

Above $900\,W/m^2$ the data become quite erratic from year to year; in 1983 a total of $18.2\,h/a$ above $920\,W/m^2$ was registered [III,3.1] although during the years 1979 to 1980 an estimate of $262\,h/a$ had been made on the basis of earlier data [TR 1/81]. Until the second half of 1984 the insolation level of $920\,W/m^2$ was never reached.

2.2.2 Winds

The mountain ranges surrounding the site affect the statistical distribution of the winds so that easterly and westerly winds predominate. The polar diagram of "wind length" shown in Fig. 5 was obtained by reporting a wind coming from a given direction if it falls within one of 24 15-degree segments [III,3.1]. The term "wind length" denotes the product of the monthly average wind speed multiplied by the length of time it blew in that particular direction.

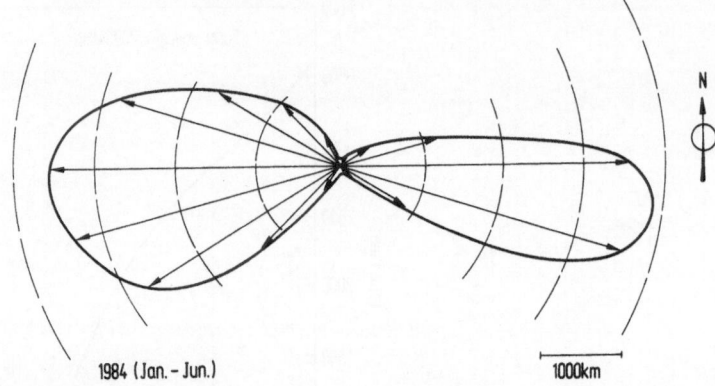

1984 (Jan. - Jun.)   1000km

Fig. 5. Wind length diagram

The observed data lead to the following conclusions:

- Westerly and easterly winds predominate.
- There are east winds every month.
- West winds may sometimes be quite strong.
- The eastern wind sector is 60° wide.
- The western wind sector is 30° wide.

The peak wind velocity observed during the first half of 1984 was 8 m/s from an east-southeasterly direction.

2.2.3 Soiling Factors

Industrial operations east in the general region, as well as the dry and sandy soil, are the main causes of dust. The dust was analyzed by V. Ruiz and J. Usero of the University of Seville [III,5.1] in order to determine the extent it would affect the reflectivity of the mirrors used to concentrate the solar energy.

The collected dust was analyzed using visible as well as atomic absorption spectrometry, sedimentation techniques and turbidity measurements. It was determined that:

- The levels of sedimentable material and of suspended particles are not high when compared to those encountered in industrial zones.
- The differences in chemical composition of the samples are very small.
- The variations of the dust level is similar in all the samples.
- Chlorides, sulphates and calcium are the major components of the soluble fraction, amounting to more than 40% of the soluble fraction.
- It follows that the amounts of sedimentable material in the fields are well represented by the average values measured at the collecting stations, which had been placed at the 4 corners of the site.

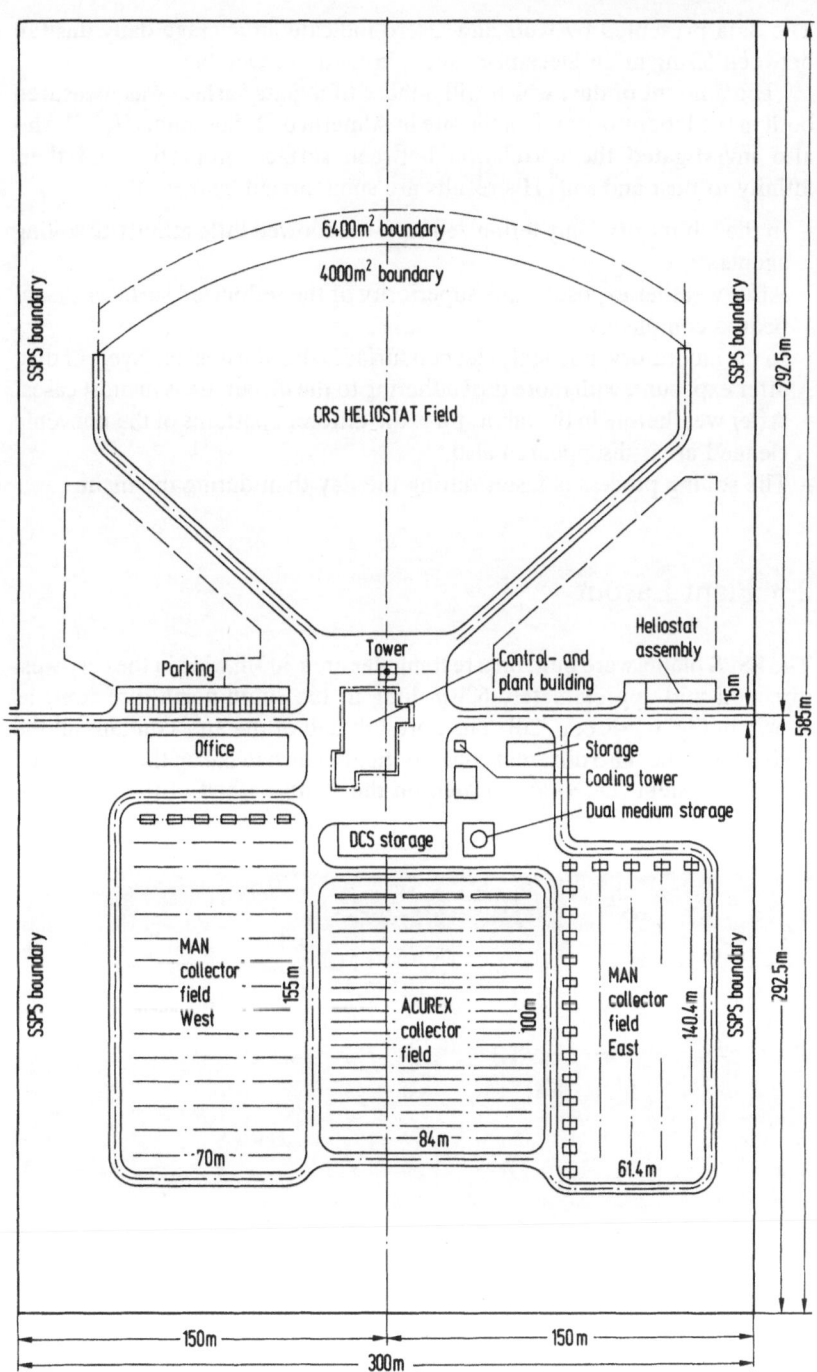

Fig. 6. SSPS plant layout

The data presented by Ruiz and Usero indicate an average daily dustfall between 63 mg/m² in December and 276 mg/m² in October.

The amount of dust which will adhere to a glass surface was measured both in the laboratory and on the site in Almería by I. Susemihl [II,5.2] who also investigated the correlation between surface properties and their affinity to dust and soil. His results are summarized below:

- In the laboratory, only teflonized surfaces showed little affinity to soiling agents.
- After weather exposure, the superiority of the teflonized surfaces disappeared completely.
- In the laboratory, unevenly cleaned surfaces showed uneven layers of dust after exposure, with more dust adhering to the dirtier areas in most cases.
- After weathering in the laboratory, the different patterns of the unevenly cleaned areas disappeared also.
- The soiling process is faster during the day than during the night.

2.3 Plant Layout

The SSPS plants were built on a rectangular area 300 m wide in the east-west direction and approximately 600 m long in the south-north direction, as shown in Fig. 6 [SR 6, p. 20]. The northern half of the area contains all the heliostats while the 3 different collector fields which make up the distributed collector system (DCS) are located on the southern half of the area.

Fig. 7. Aerial foto of SSPS

3 The Central Receiver System

3.1 General Description

As may be seen from the schematic in Fig. 8, the CRS plant consists of a field of 93 heliostats which reflect the sun onto a heat capturing device called a "receiver". The heliostats track the sun during the day keeping its image reflected onto the receiver aperture. Each heliostat consists of a number of mirrors which are adjusted individually in angle and curvature so as to obtain as small an image of the sun as economically practical. The receiver is mounted on top of a tower in order to make it possible for each heliostat to "see" the receiver at all times. The receiver absorbs the solar radiation and transmits the heat to a suitable working fluid, in this particular case, liquid sodium. The heated sodium is first pumped through a hot storage tank which provides a reserve of thermal energy for limited amount of time and also passes through a heat exchanger which is called the "steam generator"; it produces the steam required to run the power conversion system [I,2.].

It is not possible to reflect all the sunlight onto the receiver because each heliostat casts a shadow which varies in location with the time of day, thus blocking the sun from other heliostats. Near the tower, heliostats can be packed close together, but as their distance from the tower increases, their packing density has to be reduced to prevent mutual blockage. At the SSPS project the heliostats were located to the north of the tower and laid out over a field shaped roughly like a fan, such as shown in Fig. 9.

The SSPS project underwent a number of changes during the predesign stages [SR 2, see also TR 1/84], some of which were due to financial constraints. For the final design of the CRS plant, the specifications shown in Fig. 10 were laid down [SR 2]:

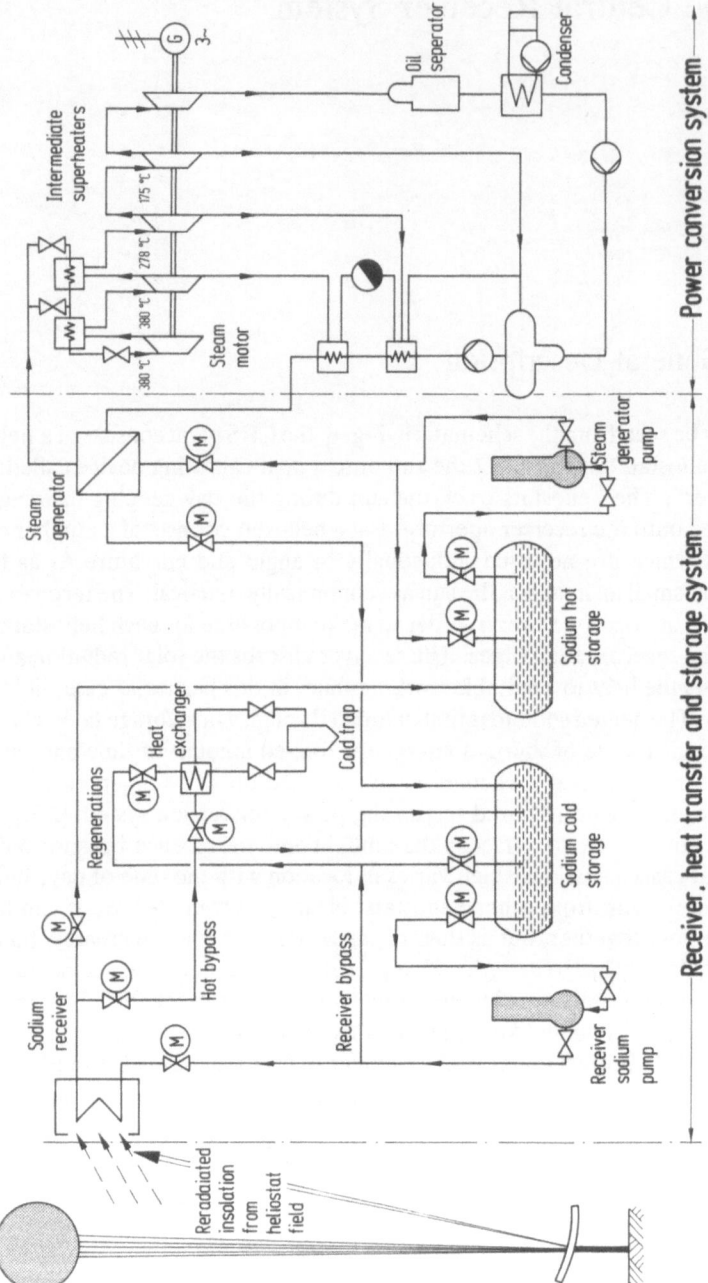

Fig. 8. Simplified schematic of central receiver system. For the sake of simplicity the field of heliostats is shown as a single mirror at the lower left of the picture

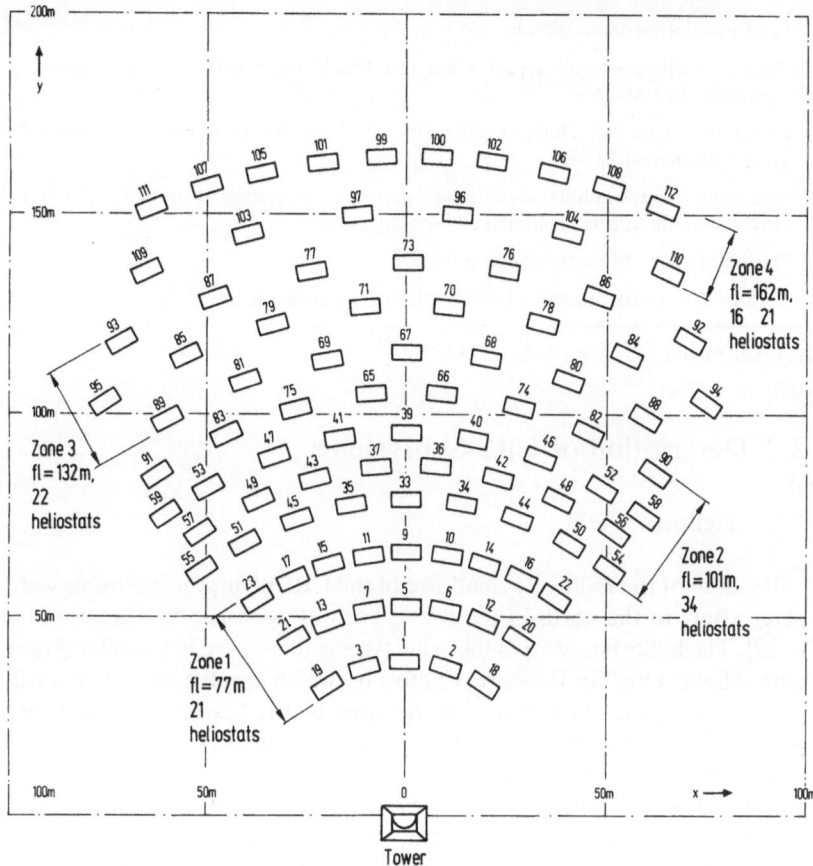

Fig. 9. Layout of heliostats and tower. The field is subdivided into 4 zones within which the heliostats are grouped by focal length of 162, 132, 101 and 77 m respectively

Maximum gross electrical output at equinox noon: 500 kWe.

Full power operation at equinox noon with an insolation of 920 W/m² measured perpendicularly to the sunbeam ("beam radiation").

Heliostat field consisting of 93 heliostats with an individual reflecting area of 39.32 m², i. e. a total reflecting area of 3657 m² resulting in a concentration ratio of 410 and a land use factor of 0.2.

Receivers:
First, a cavity receiver with an aperture of 9.7 m² in the shape of an octagon. The design peak heat flux was 0.62 MW/m² (This first receiver was designed for 160 heliostats and was later modified for the operation with the smaller number of heliostats). This receiver was later substituted by a flat receiver, the advanced sodium receiver (ASR). The ASR is a rectangular flat receiver with an aperture of 2.85 m by 2.73 m and it was designed for a peak heat flux of 1.38 MW/m².

Fig. 10. Legend see next page

Heat transfer medium: sodium.

Thermal storage system: separate hot and "cold" tanks with a storage capacity equivalent to 1 MWhe.

Power conversion: an alternator driven by a 6-piston steam motor with a calculated cycle efficiency of 27.2%.

Safety: an uninterruptible power supply as well as protection against sodium/water reactions, sodium fires, lightning and seismic events.

Design lifetime: 10 years (90 000 hrs).

Performance to be guaranteed within 90% of the design point.

Fig. 10. Main CRS design features [SR2]

3.2 Description of CRS Subsystems

3.2.1 Heliostat Field

Because of the relatively small size of the CRS plant, the heliostats were placed only to the north of the tower, such as described by Battleson [2, p. 29]. The heliostats were of the same type as those used in the solar power plant "Solar One" in Barstow, California, which is rated at 10 MW peak power ("MWp"). They were manufactured by the Martin Marietta Company and installed on site on subcontract to the Spanish company CASA. One such heliostat is shown in Fig. 11.

The heliostat mirror assembly [SR 2, p. 92–96] consists of a sandwich arrangement made by hot-bonding the glass mirrors to an aluminum honeycomb core and a steel pan enclosure. All mirror panels have the same reflective area of 1079 by 3035 m. The glass of the mirrors is 3 mm thick, the honeycomb has a thickness of 66 mm, and the plan enclosure is formed from

Fig. 11. Front view of heliostat

a single, 0.58 mm thick cold rolled steel sheet. The front surface of the mirrors is approximately flush with the edge of the pans. An expanding foam adhesive is placed around the edges of the aluminum core and the mirror assembly is finally sealed with a two-component glue. A sketch of the construction is shown in Fig. 12.

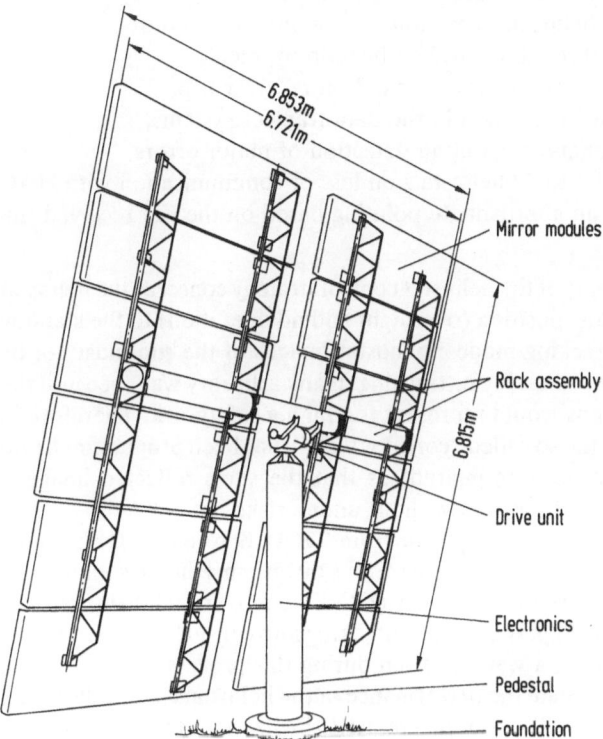

Fig. 12. Heliostat construction

The heliostat field is controlled by a computer system which uses the principle of "distributed intelligence" [SR 2, p. 99+100]:

The heliostat array controller (HAC) is a minicomputer located in the plant control room. It calculates the position of the sun every second and transmits appropriate information to four heliostat field controllers (HFC) located among the heliostats in the field. These HFC's verify the commands and transmit them, appropriately formatted, to the individual heliostat controllers (HC). These are microprocessors located in the pedestals of the heliostats which initiate motor action necessary to obtain the individual position needed by each heliostat. Two rates of control, slow and fast, are available.

The control equipment and software provide all necessary control functions for a field of up to 256 heliostats (8 HFC's with a maximum of 32 HC's), although only 93 heliostats are included in the SSPS setup. The heliostat control system is capable of a multitude of different functions, such as:

- control of the heliostats by group, by row or individually,
- monitor the operational status of the heliostats,
- detect errors of heliostat operation. i. e. inoperative motors,
- maintain safety through controlled beam movement,
- transmit emergency defocus command on receiver trip,
- detect communication errors in the data-transfer systems,
- execute an automatic reset upon detection of major errors,
- control stow of up to 32 heliostats on loss of communication with HAC computer using an approximate pointing based on the last received sun vector.

A very important part of the heliostat control strategy concerns the heliostat movement from stow position (overnight and no operation) to the standby and the receiver tracking modes: heliostat images of the sun must not be ordered from stow to standby or tracking in any arbitrary way, because the haphazard reflections would represent too high a safety risk. Therefore, a special procedure, the so called "corridor walk", has been programmed into the system. This procedure guarantees that the sun's reflected image is brought to the desired position without undue risks.

Heliostats are automatically sent from TRACK to STANDBY by an interlock system whenever there is a serious disturbance in the sodium heat transfer system. In case of emergency (DIVE command), all heliostats are directed to STOW at high speed without taking into consideration any safety corridor but activating a warning siren during this process.

In order to investigate the performance of the heliostats, a so called Flux Analyzing System (FAS) developed by EIR (Switzerland) was first used to be succeeded in later stages by a system called HERMES, developed by DFVLR (Germany) and briefly described in section 3.3.3. These systems made it possible to measure the quality of the image projected by the heliostats so that the necessary corrections can be applied to the individual heliostats.

3.2.2 Receivers

3.2.2.1 Cavity Receiver

At first, a cavity receiver designed by Interatom (Germany) and manufactured by the Sulzer Company (Switzerland) [I,5.3] was used. It had an octogonal aperture and was designed to operate with 160 heliostats. The main characteristics of this receiver are listed in Fig. 13, and photographs are shown in Figs. 15 and 16.

The absorbing surface consists of a curved wall with a mean radius of 2.25 m, a height of 3607 m and an active absorbing angle of 120°. Six parallel tubes with an outer diameter of 38 mm and a wall thickness of 1.5 mm, each

external dimensions, height = 6 m, width = 6 m, depth = 3.1 m	
aperture crossection:	9.7 m^2
active heat transfer surface:	17.0 m^2 tubes including gaps
	15.0 m^2 tubes only
inlet temperature:	270 °C
outlet temperature:	530 °C
pressure of sodium:	4 bar
sodium mass flow:	7.34 kg/s
pressure drop:	0.45 bar
incoming radiation energy at design point:	2.84 MW thermal
peak radiation density:	0.60 MW/m^2 thermal
mean radiation density:	0.16 MW/m^2 thermal
calculated efficiency:	88.3%

Fig. 13. Main characteristics of cavity receiver [I,5.3]

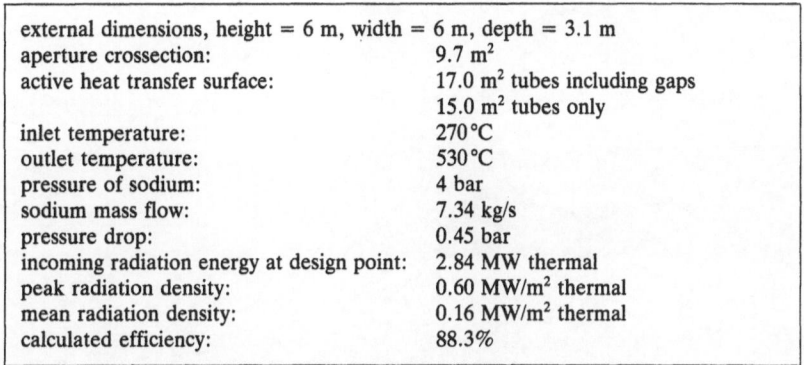

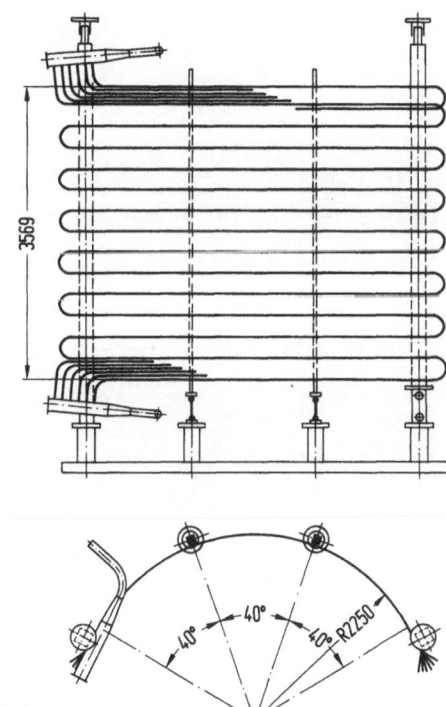

Fig. 14. Cavity receiver, absorber layout

Fig. 15. Cavity receiver, front view

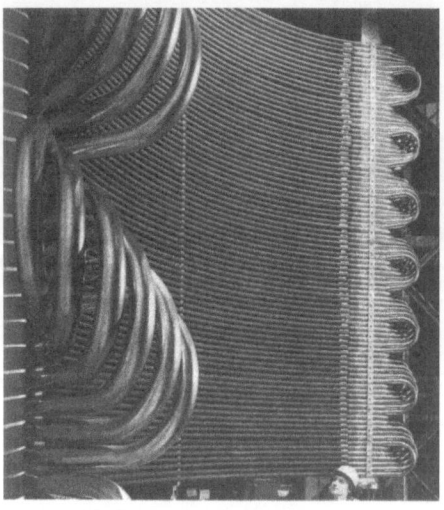

Fig. 16. Cavity receiver, inside view

87 m long, make 14 horizontal passes across the back wall of the cavity. The tubes are fixed on the one side and are free to expand vertically and horizontally.

The absorber tubes are coated with black "Pyromark 2500" paint. The door frame and the doors are coated with white Pyromark paint to protect them against the effects of "spilled" radiation. A ceramic wall behind the tubes absorbs approximately 5% of the radiated energy and was intended to reradiate it onto the rear side of the absorber tubes. The casing is airtight and can be closed by means of two sliding doors. A 350 mm thick thermal insulation limits heat losses through the casing to less than 20 kW with the doors closed.

Since this was the first receiver of such a size using sodium as a heat transfer fluid, a conservative design philosophy was adopted and extreme care was used in its construction.

3.2.2.2 Advanced Sodium Receiver

This receiver (ASR) was designed by SNAMPROGETTI (Italy) and manufactured by the Franco-Tosi Industriale [I,5.2] company (Italy) as a flat configuration of five absorber panels, resembling a billboard. Each panel onsists of 39 parallel tubes, 14 mm in diameter and 1 mm thick. (Fig. 18) Its main characteristics are listed in Fig. 17:

aperture crossection	7.94 m²
active heat transfer surface	7.94 m²
inlet temperature	270 °C
outlet temperature	530 °C
sodium pressure	6 bar
sodium mass flow	7.3 kg/s
pressure drop	1.2 bar
incoming radiation energy at design point	2.84 MW thermal
peak flux radiation density	1.38 MW/m² thermal
mean flux radiation density	0.35 MW/m² thermal
calculated efficiency	88%

Fig. 17. Main characteristics of advanced sodium receiver [I,5.2]

Figure 18 shows schematically an inside view of the absorbers and the path taken through them by the sodium. A system of support plates welded onto the absorber tubes is used to hold the absorber to the back structure. Sufficient clearance is provided to allow for thermal expansion. The ceramic wall, which consists of a double layer of overlapping ceramic tiles capable of withstanding temperatures up to 1850 °C, is situated 45 mm behind the absorber tubes. Absorber deformation was anticipated to cause gaps of up to 3 mm between tubes, thus resulting in peak ceramic surface temperatures of 1200 °C. To protect the supporting structure from excessive heating, the ceramic wall is backed by 175 mm of ceramic fiber insulation. To the left and to the right of the aperture, there are two sliding doors which can be closed to protect the receiver.

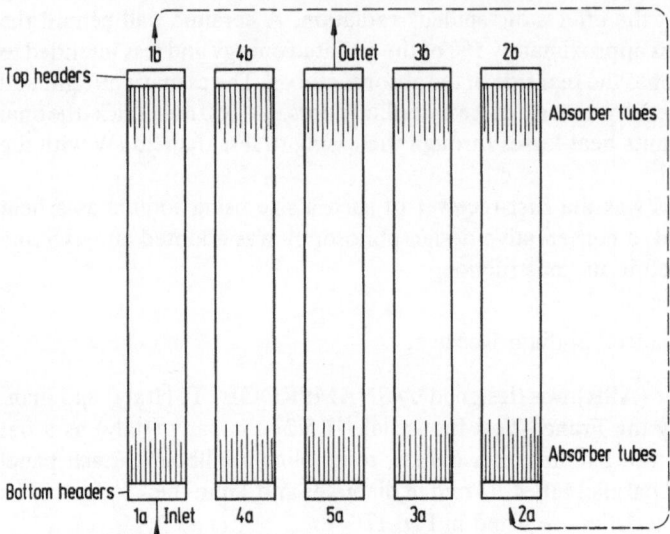

Fig. 18. Schematic view of the advanced sodium receiver

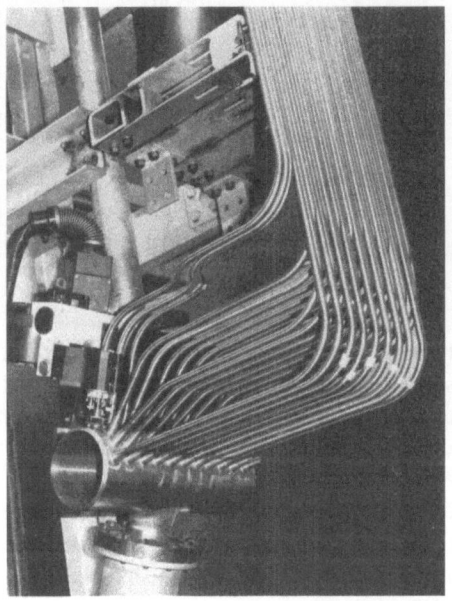

Fig. 19. ASR absorber tubes joined to bottom header

3.2.3 Sodium Heat Transfer System

The sodium heat transfer system (SHTS) transfers heat from the receiver to the thermal storage system and to the power conversion system. The thermal requirements of the CRS are:

- Receiver radiative power input: 2840 kW thermal
- Receiver thermal power output: 2508 kW thermal
- Steam generator power: 2203 kW thermal
- Receiver peak flux density: 0.6 MW/m² thermal
- Thermal storage: 2h, equivalent to 1 MWh electrical ("MWhe").

A hot and a cold storage tank, each with a capacity of 70 m, are provided. The SHTS is designed as a single-circuit system without intermediate heat exchangers but with two hydraulically independent loops: one for the receiver and the other for the steam generator. A simplified schematic of the SHTS is shown in Fig. 20.

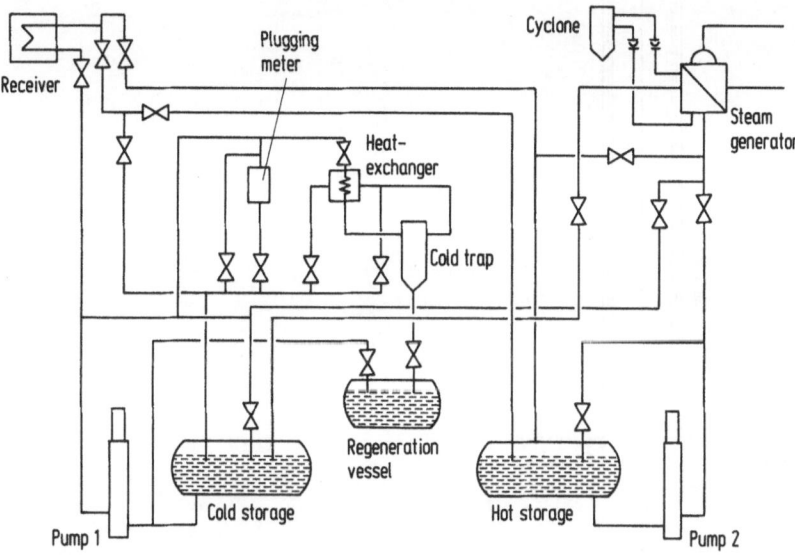

Fig. 20. Schematic of the sodium heat transfer system

Under normal operating conditions, there are two nearly constant sodium temperature levels: approximately 530 °C is maintained at the receiver outlet, at the hot storage vessel, at the hot storage pump and at the steam generator inlet; approximately 270 °C is maintained at the steam generator outlet, the cold storage vessel, the cold storage pump and the receiver inlet. All components of the SHTS in contact with the higher temperature level are made of austenitic steel, while those normally in contact with temperatures lower than 270 °C are made of ferritic steel (for example cold storage, cold trap, regeneration vessel etc.).

3.2.3.1 Steam Generator

The steam generator transfers heat from the sodium to water. It is shown schematically in Fig. 21.

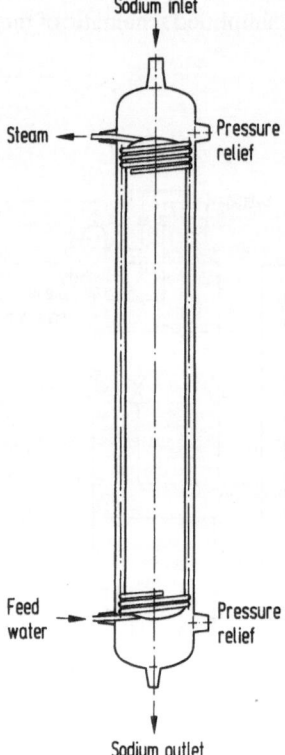

Fig. 21. Steam generator

It is a once-through, helical tube type heat exchanger. The sodium flows downwards between the outer shell and an inner displacement body around the heating tubes. The water flows upwards through the heating tubes becoming steam in the process. Two rupture discs, connected with pipes to a cyclone, are provided to prevent major damage from the consequences of possible sodium-water reactions.

The main characteristics of the steam generator are listed in Fig. 22.

Heat transfer area	14.7 m²
Dimensions of heat transfer tubes	25 mm × 3.2 mm
Tube lenght (3 tubes)	62.2 m
Thermal input power	2203 kW
Feed water temperature	193 °C
Outlet steam temperature	500–525 °C
Outlet pressure	105 bar
Steam flow at design point	0.86 kg/s
Pressure drop water/steam side	10 bar
Sodium inlet temperature	525 °C
Sodium outlet temperature	275 °C
Max. pressure on sodium side	8.0 bar
Sodium mass flow at design point	6.9 kg/s
Water/steam volume	50 l
Load range	25–100%
Overland capacity	110%
Load change rate	5%/min

Fig. 22. Main characteristics of steam generator [I,3.3]

3.2.3.2 Sodium Pumps

	Receiver loop	Steam generator loop
Normal operating temperature	275 °C	530 °C
Max. operating temperature	530 °C	530 °C
Max. flow rate	48 m³/h	35 m³/h
Max. pressure difference	3.5 bar	1.8 bar
Design flow rate	30 m³/h	30 m³/h
Flow rate range	10–155%	10–110%
Power rating	4.3 kWe	2.8 kWe

Fig. 23. Main characteristics of sodium pumps [SR2]

Both centrifugal pumps are located close to the storage vessels, one to feed the receiver, the other to supply the steam generator with hot sodium. Sodium overflow and cover gas (argon) pipes are connected directly to the vessels provided for their storage.

A cutaway view of one of the sodium pumps is shown in Fig. 24.

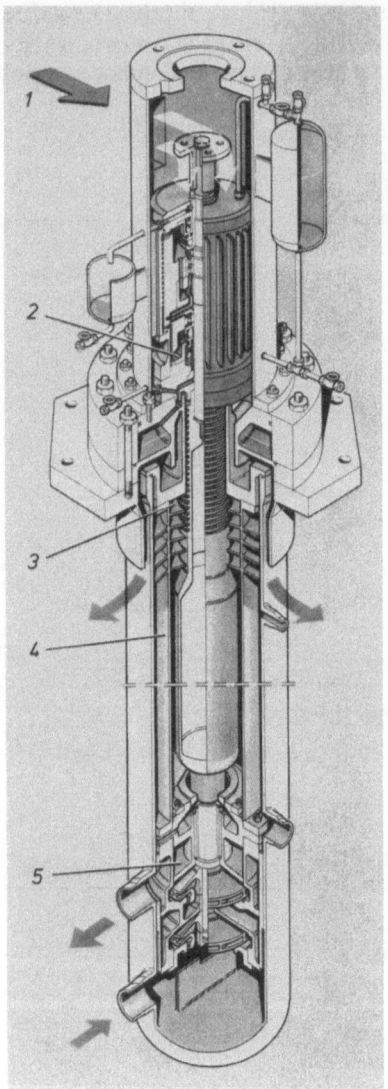

Fig. 24. Sodium pump, cutaway view

3.2.3.3 Sodium Storage Tanks

Austenitic stainless steel is used for the hot storage tank and ferritic steel for the cold sodium tank. Their thermal capacity under design conditions is 5 MWh, to provide a necessary reserve against cloud passage and other short interruptions to operation.

Fig. 25. View of the hot and cold sodium storage tanks

3.2.3.4 Auxiliary Equipment

Special mention must be made of the trace heating system, which is necessary to melt the sodium before startup and to keep it in a molten state when solar power is absent for long periods of time. Basically, the trace heating system consists of electrical heaters surrounding the sodium carrying components. The heaters are thermally insulated to minimize losses to the environment. They have an installed power rating of 135 kWe and an average power consumption of 12.8 kWe. This system is capable of preheating the entire SHTS up to approximately 200 °C.

The SHTS is controlled by three independent loops.

- The receiver loop:
  the receiver outlet temperature is used as a reference and is controlled by adjusting the speed of the receiver loop pump. A follow up controller enables this loop to adapt to fast changes.
- The steam generator loop:
  at the steam generator outlet, the sodium temperature is used as the control variable while the ratio of sodium flow to feedwater flow acts as a control variable for the speed of the steam generator pump.
- The sodium purification loop:
  within the sodium purification system, the flow rate, the cold trap inlet temperature and the temperature of purification are controlled.

A sodium purification system, a system for the protection of the free surfaces of the sodium using argon as a cover gas as well as numerous safety devices were also provided. Other safety devices, interlocks, fire extinguishing systems and many more subsystems of conventional nature were also installed.

3.2.4 Power Conversion System

Thermal power is converted to electricity by a steam motor powering a conventional alternator. The decision to use a steam motor as the drive unit rather than a turbine, was made mainly for financial reasons and also because at this power level, a turbine and its associated reduction gear was not expected to have as good an efficiency. The main features of the steam motor are shown in Fig. 26.

Nominal output	599 kW
Inlet temperature	500 °C
Inlet pressure	100–102 bar
Number of cylinders	6
Expansion	5 stages
Outlet pressure	0.3 bar abs.
Speed	1000 rpm
Overload capacity	10%
Operating lifetime	90 000 hours

Fig. 26. Main characteristics of the "Spilling" steam motor [SR2]

A diagram of the power conversion system is shown in Fig. 27, a drawing of the steam motor in Fig. 28, and a photograph in Fig. 29.

Water processing, water makeup, recooling equipment as well as separators, feed water pumps and associated gear are of a conventional nature and will therefore not be described.

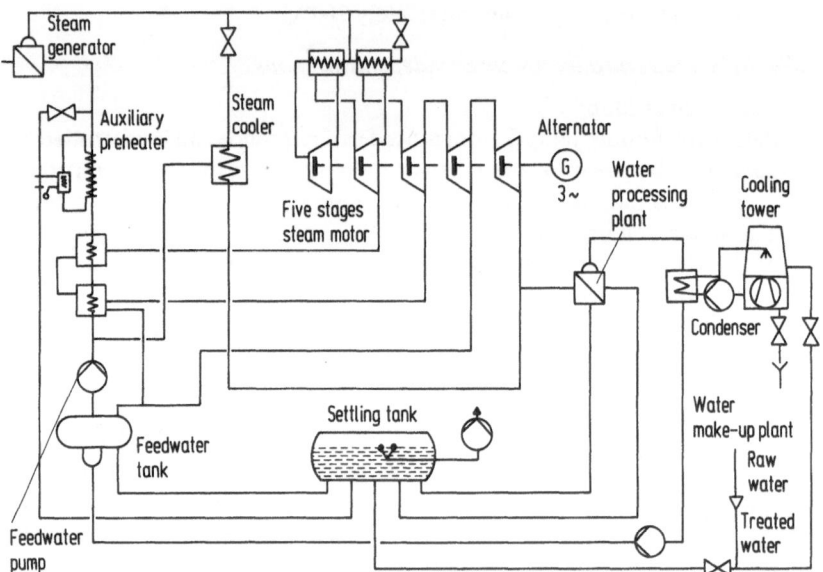

Fig. 27. Power conversion system diagram

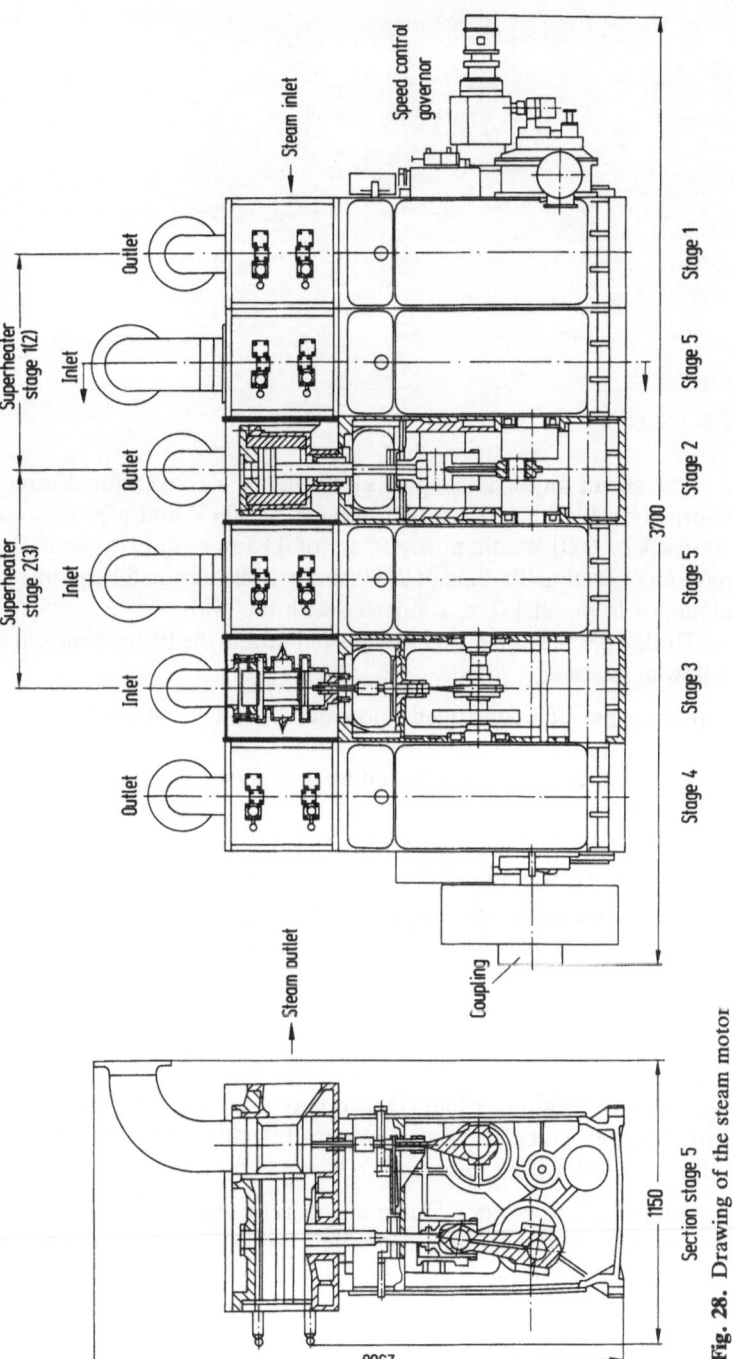

Fig. 28. Drawing of the steam motor

Fig. 29. View of the steam motor

The steam motor is coupled elastically to a conventional three phase alternator which has an output voltage of 400 V and a power rating of 700 kVA or 600 kW at a power factor of 0.85 or better. It was designed to have an operating life time of 90 000 hours. The alternator can be operated alone, or in parallel, i. e. connected with the grid.

During parallel operation with the public grid the PCS is controlled in the following manner:

– reactive power is controlled automatically and the alternator is a constant voltage type,
– the steam pressure is controlled by the speed of the thyristor controlled feed water pump,
– the inlet steam temperature is uncontrolled, but it is limited by the sodium inlet temperature of 525 °C.

In isolated operation, the output from the alternator can be varied continuously between 2 and 100% of the rated maximum output.

3.2.5 Data Acquisition System

The data acquisition system (DAS) plays a specially important role in an experimental plant of this type. It was manufactured by SAIT (Belgium) and fulfills the following tasks:

– monitoring of operators' functions thereby enabling them to make decisions on setting operating conditions (objective: plant control),
– collection and storage of all important plant data (objective: data acquisition),
– relevant information (objective: system evaluation).

The success of the SSPS test and operations phase depended very strongly on the reliability of the DAS function.

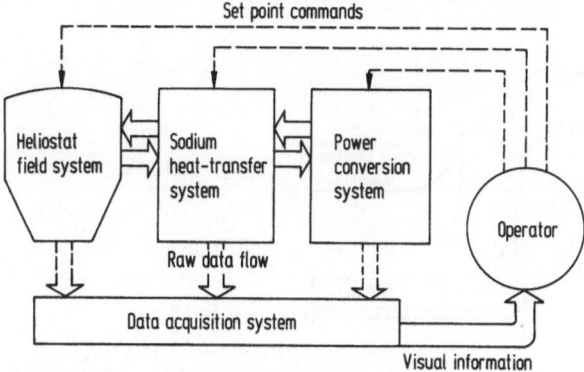

Fig. 30. Schematic of the data aquisition system

The DAS is capable of computing key results of certain test actions in real time and was therefore essential in supporting the international test and evaluation team (ITET) on the site. The hardware of the DAS is centered around a MODCOMP CLASSIC 7835 minicomputer. Its periphery, interfaces and software are described elsewhere [SR 2, p. 127–144] in great detail.

3.2.6 CRS Process Efficiency

In order to visualize the complete power generation process from solar radiation to electrical output, different types of flow charts are used by different authors; for this report the "loss tree" type of flow chart was chosen, since this is the most common presentation of the energy conversion losses among thermodynamicists.

In Fig. 31 the figures surrounded by ovals denote power in kW, those surrounded by circles indicate the conversion efficiency of each particular step and the asterisk * indicates that the "parasitic load" (explained further below) has not been subtracted. Conditions on June 7th 1984 were chosen as a typical example for a "good" day.

The number of 3125 kW shown at the top left indicates the thermal power intercepted by the heliostats operating at that particular time, i. e. the "gross thermal input" before conversion to electricity.

A total conversion efficiency of 13% is obtained if the "parasitic load", the power consumption within the plant itself, is not accounted for. Subtracting the power requirements of the plant, a net power output of 330 kW was obtained, corresponding to an efficiency of 11%. The relatively high proportion of parasitic power is a direct consequence of the small size of the power plant and its experimental nature.

Increasing the size of such plants causes their internal consumption to

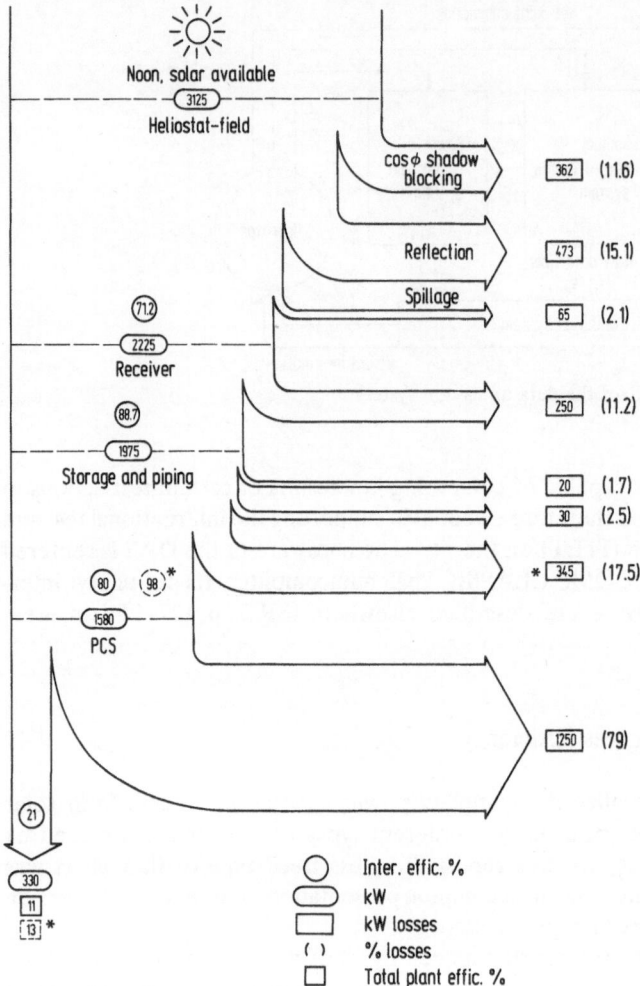

Fig. 31. CRS Power conversion loss tree (for explanations see the following text)

grow in absolute terms, but this growth is less than that of their useful output; this results in a reduction of that percentage of the gross electrical power which has to be used to cover the "parasitic load".

An operational plant would not only be larger in size and output, it would also be built without many of the subsystems needed in an experimental plant or in a demonstration system, thus the overall conversion efficiency could certainly be higher.

3.3 Measurements and Operational Experiences

3.3.1 Historical and Geographical Constraints

The original specifications called for a CRS capable of delivering 500 kWe net output whenever the solar input was 700 W/m^2 (or more) with a storage capacity equivalent to 2 MWe. The original design incorporated 160 heliostats and a steam turbine. Monetary limitations at the start of construction forced a significant reduction of the number of heliostats (only 93 active heliostats could be installed) and the thermal storage, as well as a change from steam turbine to steam motor [TR 1/84]. In consequence, the original objective to generate 500 kWe with insolation above 700 W/m^2 was changed so that the specified power level would be achieved when insolation reached 920 W/m^2, as expected at mid day at the equinoxes.

As will be seen later on, these changes had a very significant impact, not only on the performance of the plant as such, but also on the operational experiences gathered: some of the goals of the project could not be confidently reached because the plant turned out to be "subcritical" regarding the insolation available at the site.

3.3.2 General Operational Experience

This section concentrates on the availability of the plant as a whole and on the major reasons for outages, without paying attention to the particular causes of subsystem failures, which will be treated later on. The SSPS CRS plant was inaugurated on September 21, 1981 with the cavity receiver installed. Between April and August of 1983, the cavity receiver was replaced by the ASR. At the time of writing this, plant operation reports from November 1981 up to August 1984 were available. A database of all available data was set up on the VAX computer in order to facilitate data search and reduction.

In order to avoid inconsistencies and misleading comparisons, the operating hours of the different subsystems as obtained from the operators records, have been defined by Gregory, Wattiez and Blanco [I,3.1] as follows:

- Heliostats: normal tracking and standby hours not including washing time.
- Receivers: tracking hours (receiver doors open) when at least one heliostat started to track pointing on the receiver.
- Steam generator: beginning at PCS startup and including all preheating operations.
- Alternator: generator operating time when insolation exceeded 300 W/m^2 on a clear sunny day, generally referred to as a "good solar day".

In order to increase confidence in the data, the DAS tapes were also analyzed. These data also were particularly useful for determining the startup times of the different subsystems. The following definitions of operating hours apply to the DAS recordings:

- Receiver flow rate: at least $5\,m^3/h$
  Receiver outlet temperature: at least $500\,°C$
- Steam generator flow rate: at least $5\,m^3/h$
  Steam generator outlet temperature: at least $460\,°C$
- Alternator power: at least $50\,kWe$

The following search criteria were established in order to select days with stable characteristics:

1. Good solar days (stability of receiver operation is more likely on completely clear days).
2. No tracking problems (HAC).
3. Receiver operation without operating problems in sodium circuit.
4. No test operations.
5. No PCS problems.
6. Electrical generator operating.

To obtain meaningful statistical averages, only days with similar or comparable boundary conditions were chosen. In order to establish on which days stable operations could have been assumed, the following combinations of criteria were used:

- 6: electrical generator operation.
- 2 + 3 + 4 + 5 + 6: alternator operation on days when neither significant problems nor tests nor bad weather occurred.
- all criteria: as above, but only on good solar days.

Outage problems

As is often found at experimental power plants, "teething problems" account for most interruptions and outages. These problems may be divided into four categories:

1. Heliostat field problems, mostly caused by the HAC,
2. Receiver and tower problems,
3. Sodium circuit problems,
4. PCS circuit problems.

If either problems 1., 2. or 3. occur, the plant cannot be operated normally. However, if only problems in category 4. occur, no electrical power can be produced, but the sodium circuit can be operated for a duration of about one day in the hope that the repair can be completed in time for normal operation the next day. Figure 32 lists all major outage problems affecting the operation of the sodium circuit.

from	to	days	cause
15. 12. 81	28. 03. 82	103	cold tank repair
29. 03. 82	17. 04. 82	19	HAC problems with CRT
26. 04. 82	28. 04. 82	3	HAC problems with computer
29. 04. 82	10. 05. 82	12	lightning damage to HAC
14. 05. 82	26. 05. 82	13	sodium pump failure
13. 07. 82	19. 07. 82	7	sodium pump valve leak
24. 07. 82	25. 07. 82	2	sodium pump (oil fire)
24. 08. 82	27. 08. 82	4	thermocouples fitted (receiver)
07. 09. 82	20. 09. 82	14	sodium tank leak
28. 09. 82	30. 09. 82	3	PCS leakage (regeneration)
08. 10. 82	06. 03. 83	149	cold tank repair, CRT blocked
09. 03. 83	13. 03. 83	4	HAC problems
29. 04. 83	24. 08. 83	117	ASR installation
06. 09. 83	30. 10. 83	54	ASR repairs
02. 11. 83	06. 11. 83	5	ASR repairs
08. 11. 83	09. 11. 83	2	ASR repairs
02. 01. 84	04. 01. 84	2	HAC problems (leap year)
03. 04. 84	16. 04. 84	16	ASR improvements
03. 05. 84	07. 05. 84	5	lightning damage to HAC
	total	534 days	

Fig. 32. Major outage problems affecting sodium circuit operation [I,3.1]

year:	1981	1982	1983	1984
days:	61	365	365	244*
outages without failures				
weekends	12	–	1	*
holidays	–	–	1	9
cloudy days	1	4	7	12
high winds	2	–	5	1
subtotals	15	4	14	*
outages caused by failures				
HAC problems	–	25	4	7
ASR installation	–	–	117	–
ASR repairs	–	–	61	–
cold tank repair	17	172	65	–
sodium pump failures	–	22	–	1
others incl. PCS	–	33	14	28
no information	10	–	–	–
all outages (days)	42 (69%)	256 (70%)	275 (75%)	121 (50%)

(*: data imcomplete at the time of writing)

Fig. 33. Summary of all outages [I,3.1]

In many instances, a number of problems overlapped in time or the outage caused by one subsystems was used to repair or adjust another unit. Also, the plant was usually not operated on holidays, weekends and on days of poor solar conditions. In order to account for all these outages, a table listing all outages is presented in Fig. 33.

In the period from April 1982 to August 1984 the heliostat system was severly affected by two lightning strikes so at least 20 days were lost before 90% of the field could be recovered. Between May 1982 and March 1983, 21 heliostats were out of service (not including the first lightning strike). From April 1983 until March 1984, 25 heliostats were out of service. Nevertheless, during the second year of operation no more than 5 heliostats were out of service at once.

The reported period, from November 1981 to August 1984, covers a total of 1035 days. An overview of the hours of operation is presented in Fig. 34.

Plant operation hours with the cavity receiver

The cavity receiver was in operation until April 1983, a total of 544 days. Figure 35 lists days and hours of operation, as well as average hours per day of operation during this period.

	days	hours	average h/d
Cavity receiver	343	1 885.4	5.50
Steam generator	244	1 626.4	6.25
Alternator	101	238.1	2.36

Fig. 34. Total hours of operation of CRS plant [I,3.1]

	days	percent	hours	average h/d
Cavity receiver	172	32%	1 005.0	5.84
Steam generator	129	24%	798.3	6.19
Alternator	55	10%	117.6	2.14

Fig. 35. CRS operation with cavity receiver

	generator operation	no problems, no tests, no bad weather	good solar days
Cavity receiver	5.88	6.54	8.57
Steam generator	6.17	5.97	7.67
Alternator	2.14	2.12	3.07

Fig. 36. Hours of operation on days with good conditions

	days	percent	hours	average h/d
ASR receiver	171	35%	880.4	5.15
Steam generator	115	23%	728.0	6.33
Alternator	46	9%	120.6	2.62

Fig. 37. CRS operation with advanced sodium receiver

The effect of days of good operating conditions have been tabulated in Fig. 36 according to the different combinations of criteria shown earlier.

	generator operation	no problems, no tests, no bad weather	good solar days
ASR receiver	5.20	5.95	8.30
Steam generator	6.58	7.05	8.28
Alternator	2.62	2.95	3.05

Fig. 38. Hours of operation on days with good conditions

Plant operating hours with the ASR receiver

The ASR was installed subsequently and is still in operation at the time of writing. The reporting period began in April 1983 and lasted until August 1984 (491 days). Operational statistics during this period may be seen in Fig. 38.

The effect of days of good operating conditions have been tabulated according to the different combinations of criteria as done previously.

Subsystem operation as a function of daytime

In order to see how the different subsystems process the energy coming in during the day, a synoptic view of the operation of all subsystems as shown by Andersson and Sandgren [I,3.2] is presented in Figs. 39 to 41.

Figures 39 and 40 clearly indicate the main design problem encountered at SSPS: The "solar multiple" (see section 6.2 for its definition) is insufficient for a power conversion system (PCS) rated at 500 kW. However, plant operators were able to develop a procedure which made it possible to operate the plant at a reduced power output of 275 kW and using stored heat to accelerate the start up procedure in the morning [SR 7]. At 275 kW, the operation of the PCS was better matched to the radiated energy available from the heliostats. In other words, for the number of heliostats present, the PCS was too large. A typical daily output diagram obtained under such conditions is shown in Fig. 41.

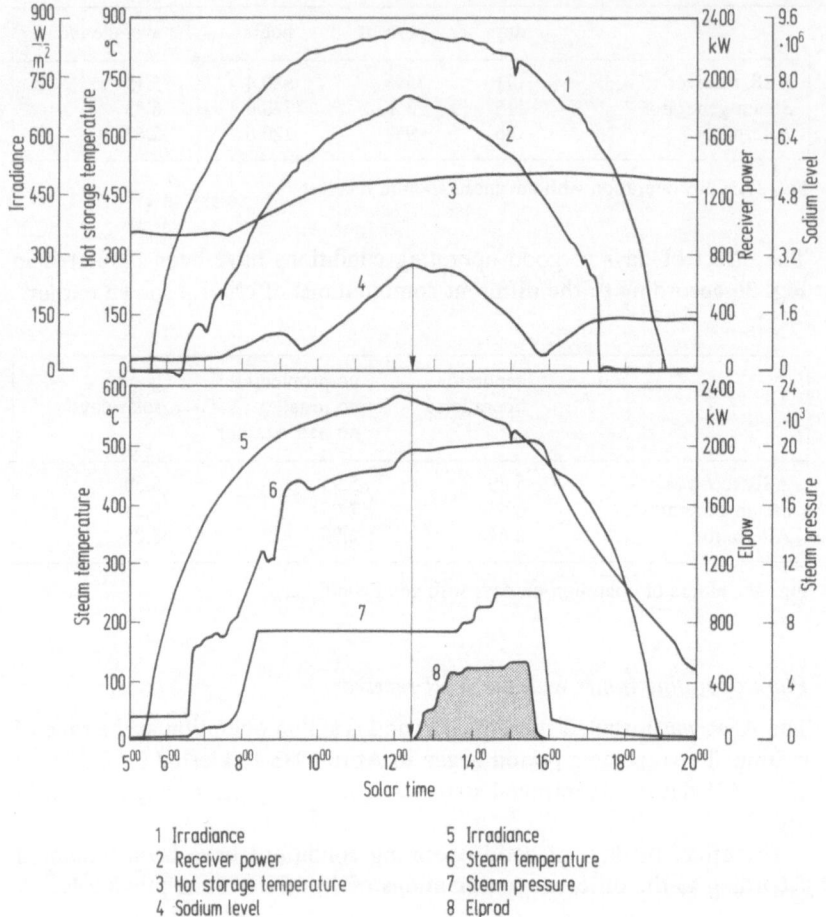

1 Irradiance   5 Irradiance
2 Receiver power   6 Steam temperature
3 Hot storage temperature 7 Steam pressure
4 Sodium level   8 Elprod

Fig. 39. Operation on July 30, 1982 (cavity receiver)

Figure 41 clearly shows that under this sort of "optimization", in which the solar multiple is "virtually" about 1.5, the plant is quite capable of generating electricity during a major portion of a sunny day. Evidently, a "real" solar multiple of 1.5 would have brought an even better capability, since the PCS was designed for 500 kW and would have been able to run at a considerably higher conversion efficiency.

Within the three years of operation (frequently interrupted by experimental work, development fixes or operational changes), this procedure could be successfully applied during 40% of good or reasonable weather. It goes without saying, that an operational plant with an adequate solar multiple would exhibit a much better yearly output.

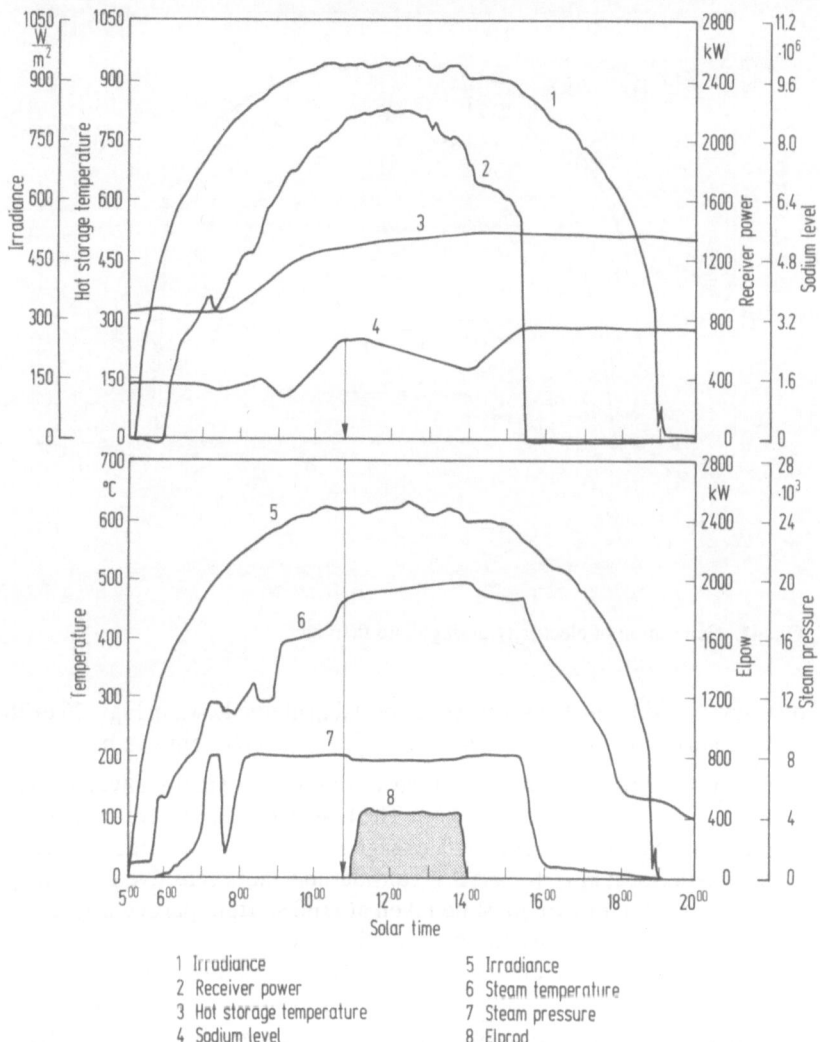

Fig. 40. Operation on June 6, 1984 (advanced sodium receiver)

Explanations and discussions

The curves from graphs 39 and 40 show that without previous storage, approximately 1.5 hours are needed to get the necessary steam quality before electricity can be generated, regardless of the type of receiver. However, the startup time can vary considerably depending on thermal energy stored, its temperature, on the receiver flow rate and the available insolation at HFS startup [I,3.2]. In particular, a startup time of 15 minutes was achieved for both receivers, when they were put into operation just after solar noon, the hot storage was at design temperature and all heliostats were

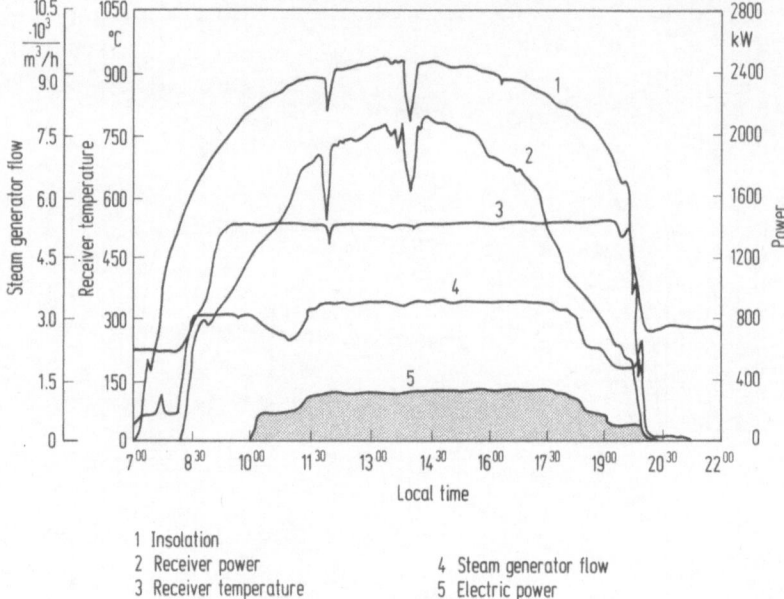

Fig. 41. Generation of electricity during more than 8h

on track. Considering the statistics of operating hours shown in Figs. 35 to 38 and the curves on Figs. 39 and 40, the following conclusions can be drawn:

– The average operating hours per day of the ASR are marginally less than those of the cavity receiver (12%). The reason for this difference may be due to weather conditions and a change in operator shift policy between the two periods; also, the ASR is considerably more sensitive to tracking errors and greater care must be taken at HFS startup, particularly in the early morning when insolation is relatively diffuse.

– Average alternator operation, particularly on good solar days, is approximately the same with both receivers; therefore generator operation is not seriously affected. This is confirmed by both the database and the CRS data tapes. Therefore it can be concluded that other factors strongly affect the behavior of the PCS.

In order to identify other factors which may influence the performance of the system in a more significant manner, the concept of "Operational Usage Factor" (OUF) was introduced [I,3.1]:

$$ \text{OUF} = \frac{\text{hours of receiver operation}}{\text{hours of insolation} > 300 \, \text{W/m}^2} $$

The figure of 300 W/m² was estimated to be the limit for the start of production of useful heat since the threshold irradiance can be calculated from the receiver total losses (about 300 kw for the cavity receiver, see

Fig. 48.). If the losses are equal or larger than the part of the incident power absorbed by the receiver, one obtains a threshold irradiance of about 170 W/m^2 at equinox 06:00 h. This value depends also on the heliostat field efficiency, which in turn depends on the day of the year, reflectance, etc. Since the receiver must be preheated, the real value for starting useful heat production is, of course, higher. Thus it was found that the OUF was larger than one when the receiver was already preheated but otherwise smaller than one, indicating that 300 W/m^2 are too low for starting the cavity receiver under normal operating conditions. For the ASR, however, this value was found to be in the range of 100 W/m^2 [2].

Summary and concluding remarks

- In 1984 outage time of the CRS plant was reduced to less than 50% of available operating time, compared with the previous 70 to 75%. These outages cannot be entirely considered as "teething troubles" of a new plant as they are almost exclusively due to major breakdowns and the long duration of the necessary repairs. In spite of this, a learning curve is evidently present.
- The outage statistics include holidays and weekends when the plant could have been operated.
- The output of the Heliostat field is insufficient for receiver operation with beam irradiances of 300 W/m^2 or less.
- In spite of considerable improvements, the electronics of the heliostat field are still too sensitive to lightning; the damage done by lightning strikes leads to long down times.
- An adequate operational strategy is needed.

One of the most important results of the experiments is that the heliostat field is too small. A larger heliostat field would lead to the following improvements:

- earlier startup time,
- faster heatup of the hot storage tank to design temperature,
- longer steam motor operation,
- increased availability factor of the heliostat field,
- increased daily system efficiency due to more energy delivered to grid.

3.3.3 Performance of the Heliostats

The daily performance of the heliostat field depends mainly on the number of heliostats able to track the receiver (their so called "availability"), on their reflectivity and on cosine – shading/blocking – effects and spillage losses [I,4.2].

Tracking was checked by closing the receiver doors and sending the heliostats individually to the "track" position. All heliostats were examined in two hour intervals at solar noon when aberration effects were minimized. When the first survey of the alignment of the heliostat field was made in December of 1983, only nine heliostats needed realignment.

In order to prevent wind forces causing excessive defocusing, the manufacturer recommended sending the heliostats to STOW whenever peak wind speeds exceeded 50 km/h. From experience it was found that the field can take higher wind speeds: since the beginning of 1984 the heliostats are stowed only when wind gusts of more than 65 km/h occur more than three times in five minutes or when a peak wind speed of more than 80 km/h is recorded. At wind speeds of more than 30 km/h the last rows of heliostats have to be defocused in order to prevent their images from overheating the receiver frame.

Besides optical errors and misalignment of the heliostats, circumsolar radiation determines the size and shape of the solar image on the receiver. Circumsolar radiation is caused by the deviation of solar rays by haze, dust or other aerosols in the atmosphere. The heliostat sees these scattered rays as if they were originating around the solar disc. Only part of the circumsolar radiation is reflected on to the receiver aperture and the rest is lost or "spilled" around the receiver opening more or less harmlessly.

A number of researchers have investigated the effects of sunshape on the solar flux at the aperture of the receiver; during 1981 and 1982 the DFVLR developed the HERMES measuring system for the investigations at the SSPS. This system consists of a special video camera in conjunction with a computer and with the DAS. It uses a sensor bar called a "target" which is capable of parallel translation across the receiver in 5 to 8 seconds. The video camera observes the receiver and supplies its digitized signal to a suitable computer which includes these data as well as those from the DAS and from the target in a real time reduction. The complete system is installed in a container with all meteorological instruments. By such means, the HERMES system makes it possible to investigate the solar shape as well as the circumsolar radiation. It was later expanded with the installation of an infrared camera which permitted more complete evaluation.

The reflectivity of the mirrors is affected by soiling and corrosion. In October 1982, after two years of exposure to the meteorological conditions encountered at the SSPS site, corrosion of the silvered back surface of the mirrors was first observed. At this time, 141 modules were affected by varying degrees of corrosion. In February 1984 the number of modules presenting some degree of corrosion had increased to 280 or 25%. The effects of corrosion were evaluated quantitatively after each survey and an exponential growth trend was observed at SSPS, as well as at "Solar One" in Barstow [I,4.3]. The conclusion was that 10% of the surfaces would be rendered useless within 10 years. At the time of writing, measures are being taken to vent the space containing the honeycomb, and the heliostats in stow are put in the vertical "wash" position whenever possible to allow condensation and rain to drip off. A new type of mirror mounting which does not use a closed space behind the individual mirrors has been developed for a follow on project.

Reduced reflectivity from soiling can be quite detrimental, because the layer of dust scatters and absorbs the incident as well as the reflected beam. In extreme cases, reflectivity can be lowered to values of 60% in 3 to 4 weeks. For this reason, washing is an important part of the maintenance of the heliostats. At SSPS, washing is accomplished frequently by a high pressure spray system mounted on a truck and using demineralized water. Once or twice a year, the heliostats are washed by hand using the classical wiping method. This is necessary since the high pressure spray is not able to restore the original reflectivity in full. The reflectivity curve for the SSPS field is shown in Fig. 42 and the effect of washing is clearly visible.

At "Solar One" the large number of heliostats (1818) has made it necessary to develop a special washing truck which wipes the entire surface of the heliostat with a brush in a very short time as shown in Fig. 43.

The availability of the heliostat field can also be adversely affected by control failures, mechanical problems, wind forces and unavoidable optical aberrations due to extreme angles of incidence of the sun.

Since the earliest days of operation, it became evident that the most important maintenance task was to keep the electronic controllers in operation. In particular, the task of repairing the controllers became quite demanding when in May 1982 lightning struck a power line which enters the plant and damaged almost 25% of them. A simple test bed for both the HC and HFC electronic cards was designed and constructed by a local electronic company so that maintenance of the controllers became faster and cheaper. Thanks to this device, after second lightning strike which damaged 75% of the controllers, it took the SSPS team only three weeks to have more than 90% of the field back in operation, in spite of the remoteness of the site and of the scarcity of resources.

When the SHTS has not been available for technical reasons or because of poor insolation, the field has been operated in a STANDBY condition,

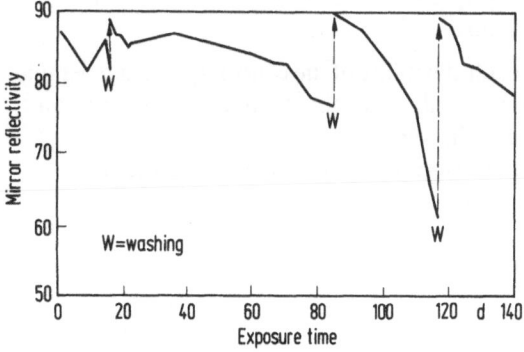

Fig. 42. Heliostat field reflectivity

Fig. 43. Solar One wash truck

i. e. with the heliostats focusing to the eastern side of the receiver, in order to gain as much experience as possible. During inactive periods, the field was sent to the WASH position (heliostats vertical) instead of the STOW position (heliostats horizontal), weather permitting, in order to reduce corrosion. Also, since the field reacts automatically to a number of special situations such as grid failure, loss of communications with the HAC etc., considerable experience was gained with the behavior of the heliostat system. The plant operators and members of the ITET have therefore formulated a number of findings [I,3.1]:

- more automation and greater flexibility of the control system is desirable,
- recovery from power system failure wastes too much operation time,
- local maintenance is a must in remote areas,
- a simpler and more cost effective washing system is a necessity,
- lightning protection is essential,
- corrosion of the reflecting surface of the mirrors may present a problem within the next 3 to 5 years.

The accuracy of heliostat performance calculations depends on knowledge of the specific heliostats in operation, on the reflectivity of the field, on atmospheric conditions, etc. A special effort was made at SSPS to obtain realistic input data for the determination of the efficiency of the receiver [I,4.2]. These measurements lead to the following conclusions:

- The measured heliostat field efficiency agrees with the calculated values.
- The measurements of the beam quality disagree with the calculated values because of the "sunshape" and because the misalignement of the heliostats was larger than expected.

Excluding failures due to lightning and to repairs of other subsystems, the availability of the heliostat field was found to vary between 96.5% and 99.5% [I,4.1], values which coincide well with those for Solar One [I,4.3].

3.3.4 Performance of the Receivers

A receiver in operation loses energy by reflecting part of the incoming light in the visible and the infrared range, by reradiating heat off its absorber surfaces, by convection (air flowing around the absorber tubes) and by thermal conduction through the structural supports. Only a typical sampling of the theoretical findings can be presented in this chapter; the reader is referred to the literature for a more complete understanding of the extensive analyses which led to the presented results. The results of the steady state investigations are presented first, followed by a description of the transient behavior of the receivers.

Since 1973 there has been extensive discussion and theoretical analysis about receiver performance, usually concentrating on receiver thermal losses [I,5.1], [I,5.2], [I,5.5]. Extensive three-dimensional mathematical models of the convection flow in and around the receiver cavity have been performed but several large experiments have led to modifications of the purely theoretical models [I,5.6–I,5.8], [41]. All these efforts have helped considerably in clarifying the different behavior of cavity receivers and flat receivers, so called "billboard" receivers. Cavity receivers are used at the French CRS "Themis", The Spanish CRS "CESA-UNO", the Japanese CRS "Sunshine", the European Community's "Eurelios" and at first at the IEA/SSPS. External receivers are in use at the American CRS "Solar One" (although cylindrical in shape rather than flat) and at the IEA/SSPS.

At the IEA/SSPS project both a cavity receiver and an external "billboard" type receiver called the ASR have been under theoretical and experimental investigation and in actual operational use. Both receiver types used the same heat transfer medium and the same heat transfer system. Some definite conclusions can therefore be drawn from the comparison between both receivers, but differences in size, difficulties with some flow measurements and with the reduction of some of the data, as well as the suboptimal orientation of both receivers in the vertical plane, call for some degree of caution in generalizing these comparisons.

3.3.4.1 Theoretical Studies and Simulations

As a result of these efforts, a number of analytical solutions for idealized problems were obtained and a method of iterative computation was developed which can be applied generally. The numerical solutions obtained for the cavity receiver and for the ASR provide a link with the work on simulation and with the results of experiments. In order to use realistic parameters for the analysis, extensive use was made of data obtained by other investigators in earlier experiments [I,5.1]. In addition to the investigation of the losses of the cavity receiver and the ASR, a receiver in the shape of a right circular cylinder made up of straight vertical tubes and surrounded by a circular field of heliostats, such as used at the Solar One CRS [I,4.3], was also investigated.

The theoretically expected efficiencies of the cavity receiver and of the ASR are plotted in Fig. 44.

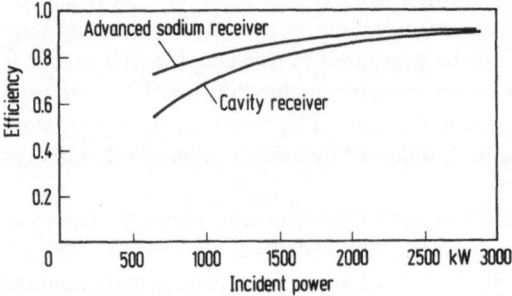

Fig. 44. Calculated efficiencies of ASR and of cavity receiver

When investigating the properties of a receiver, one must also consider its transient behavior since this determines its ability to meet a variety of operating conditions such as:

– normal and abnormal startup procedures,
– cloud passages,
– grid failures,
– experimental or accidental changes in radiation input or sodium flow.

Particular attention had to be paid to the possible effects of normal or unwanted transient phenomena; the manufacturers of the cavity receiver [I,5.5] as well as those of the ASR [I,7.3] therefore took great care to simulate these effects and to implement their conclusions in the design. Experimental and operational experiences later confirmed the design philosophies applied to both receivers.

If Fig. 45 the response functions of the two receivers have been normalized in respect to the outlet temperature obtained from simulated startups.

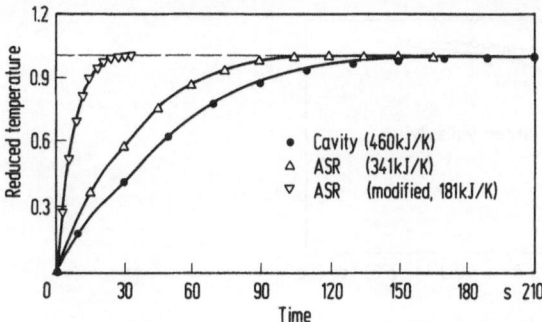

Fig. 45. Simulated transient responses of both receivers

The different response functions are the result of the different amounts of sodium in the different configurations of tubes and headers. In the case of the ASR the thermal capacity in the connecting pipes is of the same order as that of the absorber tubes. An improved ASR design using thinner connecting pipes would lead to an increased flow velocity thereby decreasing its time constant. This effect is shown in Fig. 45 (181 kJ/K) which is the result of simulating an ASR without connecting pipes between the 5 panels as a theoretical limiting case. For the limiting case the ASR would reach equilibrium after 20s while the actual ASR would do so after 120s and the cavity receiver after 150s.

3.3.4.2 Experimental Determination of Performance

Several methods were applied to determine the receiver efficiencies experimentally. An example is presented in Fig. 46 in which beam irradiance,

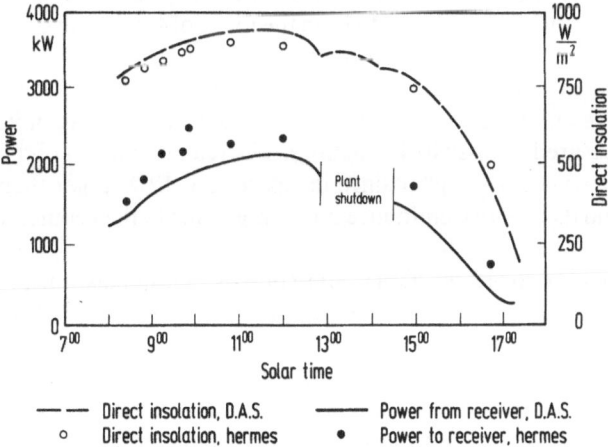

— — Direct insolation, D.A.S.   —— Power from receiver, D.A.S.
  o   Direct insolation, hermes   •   Power to receiver, hermes

Fig. 46. Insolation and power to receivers on Oct. 7. 1982

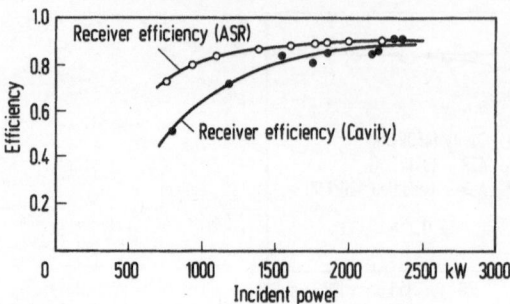

Fig. 47. Receiver efficiencies vs. incident power

(direct insolation), power to receiver (incident power) and power output from the receiver (absorbed power) are plotted as functions of solar time for the cavity receiver.

From these data and from similar data for the ASR, the efficiencies of both receivers were evaluated as a function of incident power. These plots are shown in Fig. 47.

Comparing these results with the theoretically predicted efficiencies shown in Fig. 44 it is observed that, except for low incident power on the cavity receiver, the measured efficiencies are somewhat higher than predicted.

The receiver efficiency depends on the incident power and on the following losses:

– re-radiation in the visible spectral range,
– thermal radiation losses,
– conductive losses,
– convective losses.

The re-radiation losses depend on the reflectance of the Pyromark coating on the absorber tubes, which was found to be in the range of 4 to 6%, and are therefore directly proportional to the incident power.

The thermal radiation losses depend on the mean receiver temperature and on the effective emittance of the receiver for which values between 92 and 96% were measured (emissivity of coating). The radiation losses of the cavity receiver are some 30% higher since its aperture is 19% larger than that of the ASR and its effective emittance is 6% larger due to the geometry of its aperture.

Conduction losses depend on design and construction details, on the conductivity of the materials used and on the thickness and temperature gradient of the insulating walls.

Convection losses account for the main difference in the performance of the receivers. Very sophisticated equipment is necessary to measure such losses during operation. To avoid the use of such equipment, at SSPS a method was used in which hot sodium was circulated in reverse at different temperatures without any radiative power incident on them. The convective losses determined in this manner for the ASR amounted to about half as much as for the cavity receiver.

3.3.4.3 Transient Behavior

In order to determine the transient behavior of the ASR the incident power was suddenly reduced by putting half of the heliostat field into STANDBY position. Figure 48 shows the measured values which are in good agreement with the theoretical simulation.

An example of how the cavity receiver reacts to a grid failure is presented in Fig. 49. This grid failure lead to a receiver trip. i. e. to an emergency

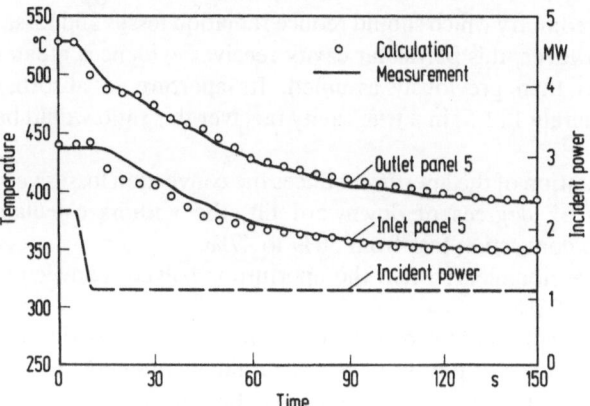

Fig. 48. Transient behavior of the ASR

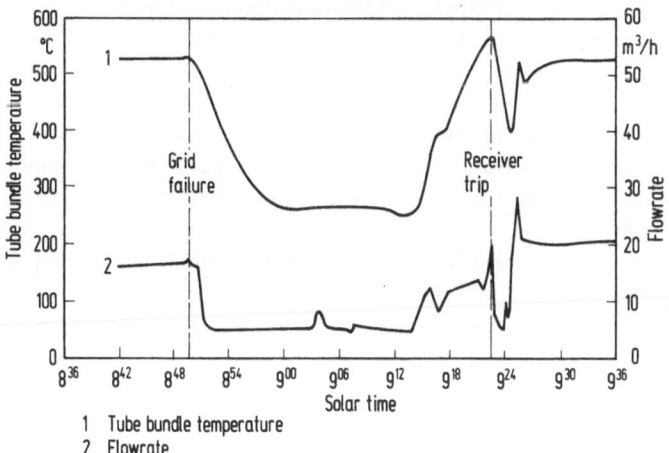

1 Tube bundle temperature
2 Flowrate

Fig. 49. Transient response of cavity receiver due to grid failure

shutdown. Since the control system did not increase the flow rate, a very high temperature reading on the absorber tubes led to the receiver trip which caused all heliostats to defocus. About 20 minutes after the failure the heliostats were put on track again.

On the basis of this occurrence, the following conclusions are drawn:

- The conservative design of the cavity receiver paid off.
- The oscillations induced by the receiver trip indicate the need for improvement of the receiver thermal control system.
- The redundancy of the system safety measures would have prevented damage to the plant in spite of the sluggishness of the receiver control.

Summarizing the results of the theoretical and experimental performance one is lead to the following conclusions:

- In principle, cavity receivers would seem to present advantages due to their protective geometry which should reduce radiation losses and losses due to wind. However, this particular cavity receiver evidences greater convection losses than previously assumed. Its aperture to absorber surface ratio is merely 1 : 1.5; in a true cavity receiver this ratio would be much smaller.
- Downward orientation of the aperture reduces the convection losses very significantly: for 32 degrees of downward tilt, the authors calculate reductions of the convection loss from 36% to 47%.
- Scaling effects are nonlinear: when the aperture is halved, convection losses go down by 23%.
- The ceramic back wall cannot produce a better temperature distribution along the circumference of the tubes as was originally intended.
- With only one tube bundle, there is no advantage in having a space between the tubes because convection losses may be increased and the receiver is made larger than necessary.
- The thermal efficiency averaged over the time of operation on a clear day is 57%; converting this to the entire day from sunrise to sunset, there would be an average efficiency of 50%.
- Response time of the cavity receiver is about 150s.
- During one of the measurement campaigns it was also found that the lowest loops of the absorber do not significantly contribute to the input; their elimination would therefore help to reduce thermal losses of the cavity receiver.
- The ASR has a high peak efficiency of about 90% and a very good part load behavior.
- The very high flux levels for which the ASR was designed could be achieved through a very compact structure.
- The ASR has a response time of about 100s.

3.3.4.4 Operational Experiences With Receivers

Although the conservative design of the cavity receiver led to a lower efficiency, it also provided some significant operational advantages: there is evidence indicating that the empty absorber tubes may have been grossly overheated during one of the usual preheating cycles resulting in a severe temperature transient while the cold sodium was filling the absorbers [I,5.4]. The damage to the surface and to the structure of the absorber is plainly visible, and yet, the cavity receiver kept working without measurable loss of efficiency and without causing operational constraints.

The back wall of the cavity receiver did not show the anticipated effect of providing short term energy storage, it merely acted as a heat shield. In addition, in the cavity receiver, the non active tube bends were exposed to the convective currents resulting in additional heat losses.

The ASR demands a fast acting receiver control system to prevent damage during situations often encountered in Almeria. However, fast acting controls are certainly within the state of the art and the quick response capability of the ASR therefore constitutes a definite advantage for the utilization of solar power in view of its stochastic nature.

As the ASR consists of a relatively large number of parallel tubes making five panels connected in series, it is necessary to monitor the tube temperatures continuously in order to avoid local overheating, especially under transient conditions and while filling the receiver. This was accomplished by a thorough stress analysis and by installing a large number of thermocouples. For this reason, a complicated procedure had to be followed very carefully when filling the ASR as explained by Ruiz and Cuadrado [I,5.10] when describing the differences between the filling strategies of both receivers.

The higher efficiency of the ASR is mainly due to the smaller heat transfer area producing smaller losses [I,5.11].

All in all, it is possible to compare the Cavity Receiver to the ASR within the framework of the purpose of the SSPS project: for the early stages of the operation of the CRS plant the Cavity Receiver offered the necessary ruggedness through its conservative design, but for the further development of the plant the ASR brought the much needed quick response capability so that the potential of the heliostat field could be best applied under the meteorological conditions encountered. In other words: heat up losses in the morning and during cloud passages could be held to a minimum because the ASR was capable of wasting little heat up time.

3.3.5 Steam Generator

The steam generator performed as designed and without cause for concern [I,3.3]. The following data were recorded:
- Inlet water temperature
- Steam temperature and pressure
- Sodium inlet and outlet temperature
- Sodium pressure
- Sodium and water flow rates
- Sodium temperature distribution

A numerical code called DISTEMP was used to perform a part load analysis. Although measurements and calculations coincided, some temperature differences were measured in the evaporator section. It was found that the heat transfer was underestimated at the film boiling condition, but this did not affect the transferred power.
The following conclusions were drawn:

- Design specifications were met and no operational problems were encountered.
- If needed, the load change rate could be improved by filling the central displacement cavity with gas.
- The present design is very adaptable to pressure and power needs.
- It would be desirable to investigate its transient behavior.

3.3.6 Sodium Heat Transfer System

When considering the operational characteristics of the SHTS, thermal losses are of utmost importance; thermal inertia is also an important constraint. Thermal losses arise from the great length of the tubing, from the large surface of the hot and "cold" sodium storage tanks and from the need for the "trace heating" system to keep the sodium in a molten state when no solar radiation is available and the sodium temperature drops below 200 C.
   Piping and tank losses were determined theoretically and compared to the measurements performed as a part of the test campaign [I,6.1]. Jacobs and Andersson derived the following conclusions, which are supported by the behavior of the system observed during normal operation.
In the storage tank subsystem the losses are shared as follows:

- roughly half the energy is lost through the tank insulation,
- roughly one third of the energy is lost through the tank supports,
- about 5% of the energy is lost through the tubing and the repair caps.

In order to show the tank losses in relation to the losses in the piping system, the pipes were numbered according to the flow diagram of the SHTS shown in Fig. 50.

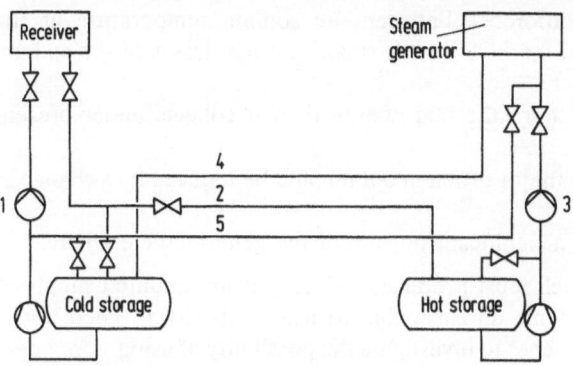

Fig. 50. CRS flowchart for the SHTS

Condition →	during tests	while collecting energy	during recirculation
Pipe 1	16.3	16	16
Pipe 2	29.3	35	17
Pipe 3	23.3	23	12
Pipe 4	–	–	12*
Pipe 5	–	–	(8)
Hot storage	23.0	23	23
Hot bypass	–	–	10*
Cold storage	17.0	17	17
Cold bypass	–	–	5*
Steam generator	–	3*	2*
Totals [kW thermal]	108.9	114	122

Fig. 51. Summary of thermal losses in the sodium heat transfer system.
(*: estimated)

Cuadrado and Wattiez investigated the losses caused by trace heating as part of the parasitic consumption [1,6.4]. They discovered that the trace heating requirements decrease during plant operation starting at values of approximately 1 000 kWh on the first day and stabilizing after 4 days of consecutive operation to a daily requirement of 500 kWh. When the plant is not in operation the trace heating requirement increases from about 600 kWh daily to over 1 800 kWh after the 4th consecutive day of no operation.

Investigation of the thermal losses, energy requirements for the trace heating system, and the effects of thermal inertia, as well as consideration of the thermal requirements of the power conversion system (see following chapter), lead to the following conclusions:

– the system is limited by thermal inertia and losses in the piping and in the storage tanks,

– as normally the difference between the sodium temperature at the receiver outlet and the PCS inlet is small, thermal losses in the sodium tanks are important,
– the PCS requires more thermal energy than it collects under present conditions,
– thermal inertia is a major problem but it could be reduced by a change in design,
– trace heating uses a significant portion of the generated electricity.

In a power plant which must produce electricity from a limited supply of thermal energy, it seems unreasonable to use electricity to keep things warm; it would make sense to investigate the possibility of using stored heat for trace heating.

3.3.7 Power Conversion System

PCS thermal losses come from three main sources [I,6.2]:

1. Losses due to low sodium temperature; energy is lost from the hot storage tank previous to startup.
2. Actual start up losses due to the requirement to warm up the steam motor and its associated gear.
3. Normal losses during operation.

From the description of the startup procedure given by Jacobs and Carmona it becomes evident that the necessity to avoid thermal shocks is in conflict with the necessity to make efficient use of the available thermal reserve; the first demands a slow increase in temperature, the second calls for as quick a start as possible. The thermal losses measured during start up from "cold" condition range from 1 200 to 1 600 kWh.

From the time when the steam motor is started, to the time when power is fed to the grid, the measurements indicate losses of 600 to 800 kWh.

Therefore, the entire PCS startup procedure costs about 20% of the energy production on a good day.

By mathematical analysis of the measured data Jacobs and Carmona arrive at the conclusion that the fixed losses, i. e. the load independent losses, amount to a constant 425 thermal kW. Load dependent losses appear to be in the order of 2.5 times the net electrical output.

3.3.8 Yearly Production of Electricity

As the CRS was used mainly for testing subsystems and measuring their performance at different power levels, and as several reasons caused frequent planned shutdowns or forced outages, the plant was not run for any prolonged period of time. An actual yearly output could therefore not be determined. To gain some insight into the power production potential, an energy loss tree for one typical day is presented. Care should be taken when extrapolating yearly energy from this instantaneous data: this is possible only if a sufficient solar multiple and the necessary storage capacity are assumed.

In Fig. 52 the numbers surrounded by ovals denote energy in thermal or electrical kW; those surrounded by circles indicate the conversion efficiency of each particular step. The net overall net efficiency of 8.1% refers to the energy actually delivered to the grid, i. e. gross electrical output minus parasitic load.

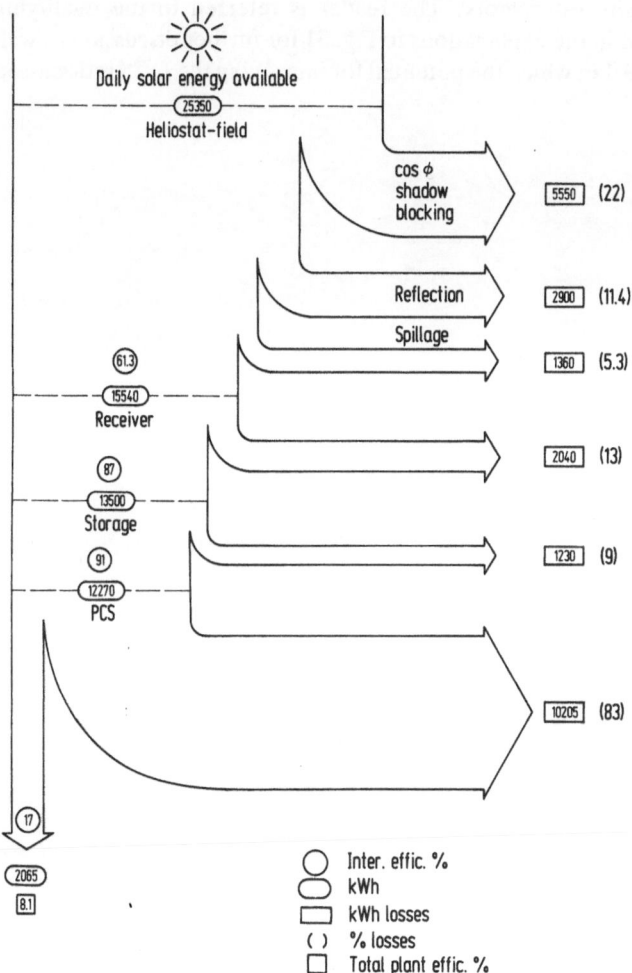

Fig. 52. Energy loss tree for one day measured an February 2nd, 1984 by Martin and Wattiez [I, 82]

The total conversion efficiency is lower than that shown in Fig. 31 for a number of reasons:

- The conversion efficiency shown in Fig. 31 refers to power conversion around noon when thermodynamic conditions are at best.
- The conversion efficiency shown in Fig. 52 refers to energy conversion (and not to power conversion!) for a whole day, during which conditions vary a great deal.
- Conditions on February 2nd were obviously worse than during a typical summer day.

Taking these considerations and the "subcritical" size of the plant into account, the net energy conversion efficiency shown for a winter day must be regarded as quite satisfactory. The reader is referred to the qualifying remarks offered in the explanations to Fig. 31 for further discussion as well as to section 6.9.1 in which the potential for improving the DCS is discussed in detail.

4 The Distributed Collector System

4.1 General Description

The energy collection components of the DCS plant are parabolic trough collectors. Sunlight is focused on a blackened absorber pipe in the focal line of the collectors and a high temperature oil which acts as a heat transfer fluid is pumped through it. In order to reduce convection losses, the absorber pipe is surrounded by a transparent tube. As can be seen in Fig. 53, the DCS can be subdivided into three subsystems: the collector fields, the storage system and the power conversion system.

The thermal oil (Santotherm-55) from the storage tank normally enters the collectors at a temperature of 225 °C and leaves the collectors at 295 °C. Hot oil is pumped from the storage tank to a steam generator to produce steam for the turbine of the power conversion system. The low temperature

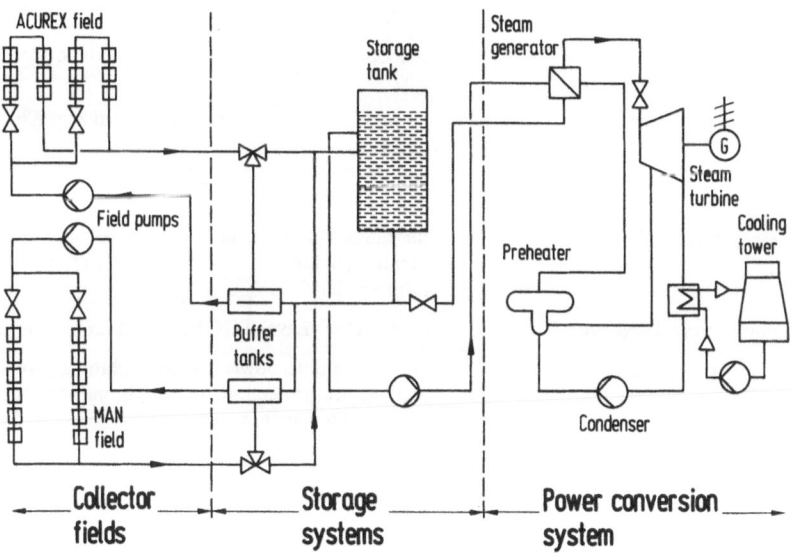

Fig. 53. Simplified diagram of distributed collector system

oil from the bottom of the main storage vessel is pumped back to the collector field.

The two types of collectors have been installed on three adjacent fields, as may be seen in Fig. 6 (see section 2.3). The center field contains the one axis tracking collectors manufactured by ACUREX (USA), the west and east fields contain the two axis tracking collectors manufactured by MAN (FRG).

The ACUREX model 3001 collectors are situated in the middle field. They intercept the sun with an active area of $2\,674\,m^2$ while covering a total of $8\,400\,m^2$ of ground.

The west field contains MAN model Helioman 3/32 full tracking collector modules which intercept $2\,688\,m^2$ of sunlight and occupy $10\,850\,m^2$ of ground surface.

Additional MAN modules were installed on the east field and intercept a total of $2\,244\,m^2$ of sunlight; they occupy an area of $8\,620\,m^2$.

A loop in the collector field is defined as an interconnected set of collectors capable of raising inlet oil temperature to the outlet temperature. There are ten loops in the ACUREX field and each loop is formed from two rows of collectors. There are 14 loops in the MAN west field with 6 collector modules per loop. A MAN east loop consists of 7 collectors arranged in two subfields with 5 loops in each.

Effective collector area Collectors	$7606\ m^2$ 14 loops double axis tracking MAN "Helioman 3/32" 10 loops improved single axis tracking ACUREX "3001" 10 loops double axis tracking MAN "Helioman 3/32"
Thermal storage	thermocline storage equivalent to 0.8 MWh electrical dual medium storage tank equivalent to 0.7 MWh thermal
Heat transfer fluid	thermo oil for plant operation at ambient temperature, below 0 °C and for startup at a temperature of 50 °C.
Power conversion drive	adapted Stal-Laval TGC 8/2500 steam turbine
Safety precautions	Uninterruptible power supply, lightning protection, fire protection and provisions against earthquake damages
Design lifetime	10 years (90 000 hours)
Guarantee	90% performance at design point upon acceptance

Fig. 54. Main DCS design characteristics [SR1]

4.2 Description of Subsystems

4.2.1 ACUREX Single Axis Collectors

This collector field consists of one axis tracking collectors oriented east west (i. e. tracking the sun in elevation only) manufactured by ACUREX (USA). The reflectors are made by GLAVERBEL (Belgium) by silvering thin glass (0.6 mm) sheets, bonding these sheets to flat steel sheets which are then elastically bent into the shape of a parabolic cylinder, a shape which they assume naturally by being mounted in an appropriate frame. The solar energy is concentrated on a receiver tube which is coated with a selective coating and covered by a transparent tube for mechanical protection and in order to reduce convection losses. A close up view of some of the single axis collectors is shown in Fig. 55. The length of an individual mirror is 3.05 m and its aperture is 1.83 m. Six such mirrors are arranged in a row which has a total length of 19.75 m and is driven by one common unit.

Fig. 55. View of a single axis collector

4.2.2 MAN Double Axis Collectors

These collectors were developed by the German company M.A.N. under the trade name "Helioman" model 3/32 and were used in several solar power projects. They are driven in a manner somewhat reminiscent of the method used on the CRS heliostats. A complete collector is called a module, the height of each module is 5.16 m and its width 7.96 m. The north-south spacing between modules is 10.5 m and the east-west spacing is 12.0 m. The

Fig. 56. Front view of a double axis tracking collector

MAN collector modules have each an aperture area of 36 m². A view of a two axis tracking collector is shown in Fig. 56.

Later on, a third collector field was added as additional MAN collectors had become available from a canceled project.

4.2.3 Heat Transfer System

The heat transfer system consists of three major loops:

- The first loop extracts low temperature (225 °C) oil from the storage tanks, circulates it through the collector fields and returns it at high temperature (275–295 °C) to the top of the same tank.
- In the second loop, a steam generator extracts the thermal energy from the oil and returns it to the thermocline tank.
- The third loop circulates water or steam through the power conversion system.

As may be gathered from these descriptions, extensive piping is required to transfer the thermal energy from the collector fields to the storage systems and to the power conversion unit.

At first thermal energy storage consisted of a single vessel using the thermocline principle: the hot oil has a natural tendency to remain in the upper section of the tank, while the cold oil stays in the lower part. As the hot oil is used, a comparable volume of cold oil is pumped into the lower tank section thus keeping the total volume change relatively small while the boundary between cold and hot oil moves upward inside the storage tank. Upon replenishment of the upper part of the tank with hot oil, the opposite process takes place moving the temperature boundary, the "thermocline", downward.

Later on, a dual medium storage tank (DMST), which had been developed for another project, was added to the DCS plant in order to gain some experience with this concept and also to supplement the existing

thermal storage capacity. This particular subsystem uses a stack of 115 slabs of cast iron to store the thermal energy. The oil circulates around the cast iron slabs as a heat transfer medium.

There are two buffer tanks which are considered part of the DCS storage system, one for the ACUREX and one for the MAN collector fields. These tanks are used to prevent cold oil from the collector fields from entering the thermocline storage vessel while the collector fields are being heated.

As with all hydrocarbon oils, slight thermal cracking occurs at elevated temperatures; this is accelerated in the presence of oxygen which also introduces a flammability hazard. An ullage and pressurization system is therefore provided which uses nitrogen in order to prevent oxygen from coming into contact with the hot oil. The ullage pressure is held slightly positive in regard to ambient so as to prevent the air from entering the system; it also accommodates changes in the fluid level due to thermal expansion and contraction.

A fluid makeup system replaces the degraded thermal oil and a catch basin surrounds the thermal storage tanks to protect the area in case of accidental rupture of the tanks.

The steam generator consists of an economizer, an evaporator with a steam/water separator mounted on top and a superheater.

Storage capacity	1500 kg
Steam output	1372 kg/h (min)
	3838 kg/h (nom)
	4230 kg/h (max)
Steam pressure	28 bar (max)
Steam temperature	285 °C
Oil temperature	295 °C (in)
	225 °C (out)
Oil mass flow	46 700 kg/h (nom)
Oil pressure drop	1 bar

Fig. 57. Main characteristics of the steam generator [SR1]

In order to be able to control the flow of the fields independently, separate pumps are provided for each collector field. To minimize spare part requirements, a single centrifugal pump design is provided. The centrifugal pump tolerates the large changes in viscosity to be expected from the large temperature variations and has high seal reliability. The pumps are able to start pumping the oil for cold startup at 5 °C.

Each pump is individually controlled on the basis of field outlet temperature and insolation; its speed is adjusted to minimize the pressure drop across the flow control valve so as to reduce the pumping losses associated with high differential pressures while matching field flow to insolation to deliver oil at constant temperature to the hot end of the storage system.

A view of the major components of the DCS heat transfer system is presented in Fig. 58. On the left the thermocline storage tank is visible as a

Fig. 58. View of storage tank, buffer tanks and hot oil pumps

large vertical cylindrical structure. At its base to the left, one of the hot oil pumps is just barely visible and both buffer tanks can be seen towards the right edge of the picture.

4.2.4 Power Conversion System

The power conversion system (PCS) consists of an eight stage condensing turbine (7 stages with one extraction) made by STAL LAVAL (Sweden)

Turbine inlet pressure	25 bar
Exhaust pressure	0.07 bar
Speed	9962 rpm
Output shaft speed	1500 rpm
Calculated efficiency, turbine + gear	23.9%
Alternator rating	713 kVA
Max. output at 0.8 power factor	577 kWe
Efficiency at 0.8 power factor	95%
Voltage	400 V/3phase

Fig. 59. Main characteristics of DCS power conversion system

which drives an air cooled, 4 pole alternator through a reduction gear. The generator and the turbine are lubricated by a common system. The generator windings are capable of withstanding short circuits.

The electric system was built by TECNICAS REUNIDAS (SPAIN). It permits connecting the PCS to the local grid through a transformer in the conventional manner. For simulations and for plant stand-alone operation, a substitute resistive load is provided which can be used jointly with the CRS plant. For safety, an uninterruptible power supply and an emergency diesel power group is provided. This is also necessary to start the DCS plant in case of local grid failure.

4.2.5 Data Acquisition System

As in the case of the CRS, the data acquisition system of the DCS converts and stores the measured data and handles and stores the computational procedures needed to evaluate the performance of the plant. The computer based system is subdivided into two main functions, the master control and the data acquisition. It was installed by ELEKTROWATT (Switzerland) to

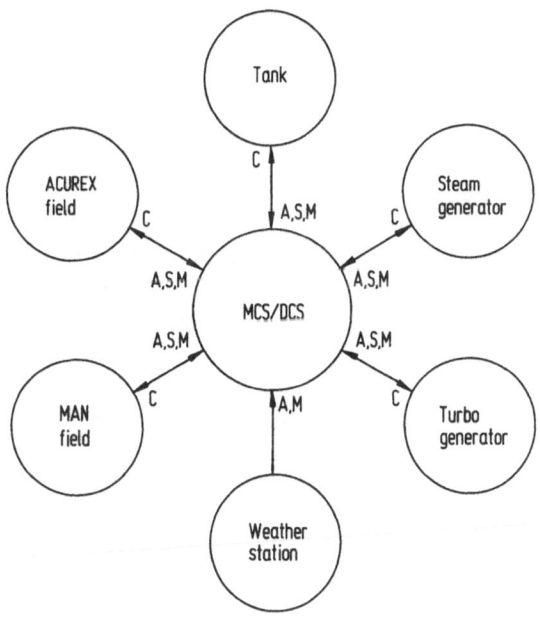

C Command signals   S Status signals
A Alarm signals   M Measurement signals

Fig. 60. Interface between MCS + DCS and plant

supervise the subsystem controllers (which can operate each plant independently, i. e. without the master control system) and to provide alarm messages. In addition, the DAS can detect and indicate malfunctions of the subsystem controllers and control automatically all operational modes, except for some startup and shutdown procedures. Special attention was given to the ability of recording alarm histories.

The DAS produces printouts and colored graphic displays. The computer used for master control functions and for DAS is a Hewlett Packard 1000/ Model 40 with a 256 kByte standard performance main memory. The DCS software is written in ANSI standard FORTRAN IV. The main disk stores data from alarms, plant log messages and all raw measurements from the last one hour of plant operation. Essential inputs can be transferred to a file large enough for ten days of operation, afterwards the data are transferred to tapes. The block diagram in Fig. 60 shows how the DAS interfaces with the plant subsystems.

4.2.6 DCS Process Efficiency

The efficiency of the energy conversion process from solar radiation to net electrical power as anticipated in the design specifications was presented by Wattiez as a flow chart from which Fig. 61 was derived.

4.3 Measurements and Operational Experiences

4.3.1 Historical Development

As is the case of the CRS, insolation data were overestimated from unreliable statistics obtained earlier and financial limitations made it necessary to reduce the number of collectors during the early stages of construction. However, the effect of this reduction was offset in part by the installation of the third collector field, using an improved layout for the new double axis tracking collectors, and by the addition of the dual medium storage tank.

As the DCS is relatively uncomplicated and as the three collector fields can operate independently, the plant could be kept in operation in a reduced state whenever this appeared beneficial for experimental or other reasons, only the general behavior of the plant is presented. Individual experiences with the different subsystems are also presented whenever significant.

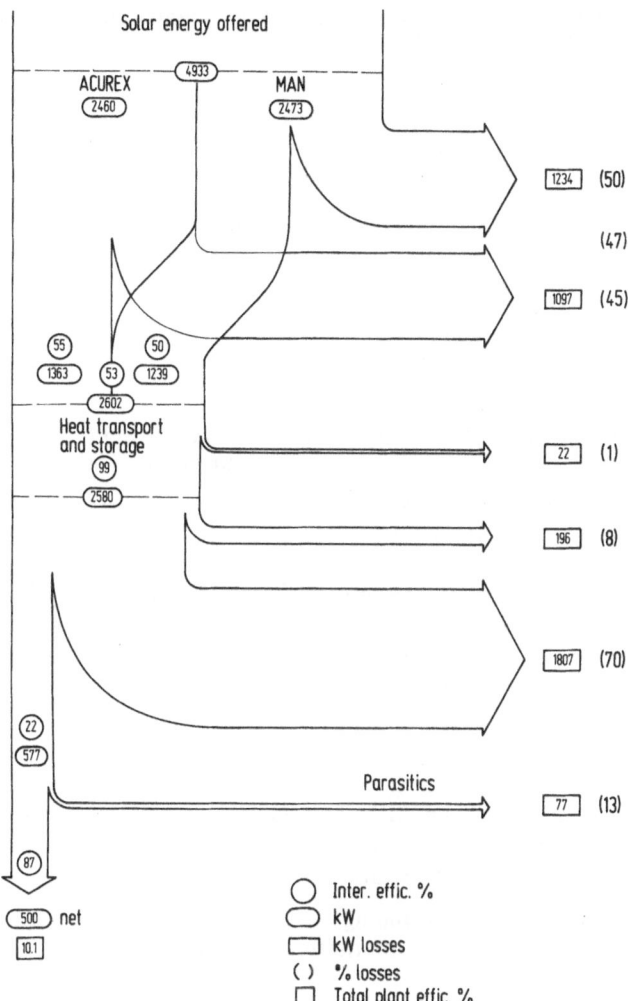

Fig. 61. DCS power conversion loss tree as calculated for the design specifications [II, 3.1]

4.3.2 General Operational Experiences

1. *First year of operation*

Wattiez [II,3.1] presents the average operational characteristics of the DCS for the first year of operation, from October 1981 to September 1982, as shown in Fig. 62. The plant reached functional adjustment at the end of May of 1982. However, there was frequent need for additional adjustments throughout the remainder of the year. The DCS plant was shut down 22% of the time purely because of bad weather or insufficient insolation, 26% of the remaining operational time was devoted to routine maintenance and to improving the system:

	operating time, h	utilization factor	thermal efficiency	conversion efficiency
ACUREX field	1135	0.51	38.6%	–
Man field	1542	0.74	35.3%	–
PCS	460	0.67	–	18%

Fig. 62. First year average performance of the DCS (October 1981 to September 1982), [II,3.1]

For the collector fields, the utilization factor is the ratio of hours of field operation to hours when beam irradiance is above 300 W/m².

For the PCS the utilization factor is the ratio of gross electrical energy produced to the product of nominal power (577 W) and time the alternator was connected to the grid.

The collector field efficiency is the ratio of thermal energy produced to solar energy available.

The PCS efficiency is the ratio of electrical energy produced to thermal energy delivered to the PCS.

2. Second year of operation

The second year was more representative of normal plant operation, as most of the detailed measurements connected to the acceptance phase had been concluded. The insolation statistics were:

- 40% good solar days (clear and shining),
- 38% medium solar days (hazy or intermittent clouds),
- 21% poor days (bad weather),
- 2820 hours of beam radiance above 300 W/m²,
  (predicted value: 3000 hours).

Total outages for technical reasons kept the plant down for 5% of the operable time and the plant was operated without producing electricity for 7% of the time. The major technical causes for outages were connected to malfunctions of the master control/DAS, which kept the plant inoperative for 10 days and caused partial or total loss of data on 46 days. In addition, there were failures in the feedwater pump, turbine speed regulation and alternator synchronisation. The plant also had to be shut down for 4 days in order to add a supplement to the DCS system. The need to maintain the reflectivity of the mirrors at an operable value caused a 9 day down time of the ACUREX field and prevented the MAN fields from being operated during 10 days.

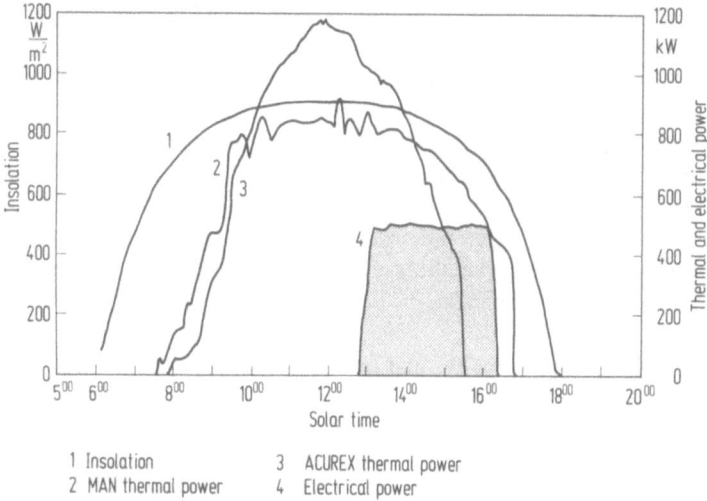

1 Insolation   3 ACUREX thermal power
2 MAN thermal power   4 Electrical power

Fig. 63. Typical DCS operation on March 18, 1983

3. *Third year of operation*

From October 1983 to September 1984 the plant operated for its third year. In March 1984 the third collector field was added bringing the total collecting area from $5362\,m^2$ to $7606\,m^2$. During the third year the weather offered:

– 34% good solar days (clear and shining),
– 40% medium solar days (hazy or intermittent clouds),
– 26% poor solar days (bad weather),
– 2627 hours of beam radiance above $300\,W/m^2$.

During the third year, the major shutdowns were caused by the Master Control System / Data Aqusition System which prevented the plant from operating on 19 days and caused loss of data on 23 days. Steam generator leakages rendered the plant non productive in terms of electrical generation for 13 days.

Other equipment failures had little effect on the operation of the DCS plant, but the lack of personnel and shutdowns due to weekends and holidays put the plant out of service for 82 days, or 22% of the year.

4.3.3 Performance of the Collector Fields

4.3.3.1 Availability and Outages

During 1983 and 1984, Schreitmuller [II,3.2], and Swanson + Fazzolare [II,6.2] collected operational data and performed evaluations which help in understanding the major causes of down time.

Availability of the system was defined as hours in which beam irradiance exceeded 300 W/m. With this ground rule, and excluding weekends and holidays, the following availabilities can be computed from the data of Swanson and Fazzolare. In Fig. 64, sng denotes the single axis collector field and dbl both double axis tracking fields.

Although three years may seem a short time from a statistical viewpoint, the POA was able to operate the DCS plant 7 days a week from the manual in a routine manner so that the statistical data can be considered quite meaningful as far as "normal" availability of the collector fields is concerned.

The outage hours of the collector fields are tabulated in Fig. 65 for the same period of time.

The man hour requirements to keep the fields operating have been summarized by the same authors, see Fig. 66.

Month	total sun hours/mo.	insol. >300 W/m² minus holidays and weekends	hours of operation		availability	
			sng	dbl	sng	dbl
Jan	302.65	109.62	88.42	85.10	.807	.776
Feb	276.43	98.74	71.01	69.58	.719	.705
Mar	364.33	146.90	104.81	78.13	.714	.532
Apr	389.06	111.18	89.33	98.08	.804	.882
May	433.63	139.03	108.44	123.39	.779	.888
Jun	391.55	149.69	83.73	109.57	.559	.732
Jul	442.12	197.20	153.85	187.59	.780	.951
Aug	401.20	201.15	99.97	103.67	.497	.515
Averages					.707	.748

Fig. 64. Availability of DCS during 8 months in 1984 [II,6.2]

Month	single axis	dual axis
Jan	0.00	59.02
Feb	0.87	18.50
Mar	4.83	20.55
Apr	0.00	42.00
May	0.00	37.67
Jun	0.00	24.88
Jul	10.05	10.87
Aug	0.00	1.92
Total (hours)	15.75	223.48

Fig. 65. Outage hours during first 8 months of 1984

Month	single axis	double axis
Jan	0	220
Feb	69	127
Mar	18	129
Apr	3	57.5
May	10	93
Jun	44	124
Jul	24	14
Aug	9	40
Total (hours)	177	804.5

Fig. 66. Man hour requirements for the collector fields

The single axis tracking modules installed at first presented a particular problem; although it has not yet affected the efficiency of the reflectors or the operation of the DCS, it is worth mentioning as a degradation of a typically «solar specific» component.

Jacob, Mertens and Declerk [II,6.4] investigated a phenomenon called "mirror delamination". The bonding agent used in the first generation single axis mirrors to glue the thin glass to the steel sheet had poor long term resistance to high temperature and to frequent temperature changes. The delamination was visible on a number of mirrors and investigations were made with strain gages and optical measuring devices to observe the conditions under which delamination would occur. After concluding the measurements and some mathematical modeling by the manufacturer in June 1983, 127 panels were replaced by new generation mirrors which had been bonded to the steel base with twice the previous thickness of a different bonding agent. This solved the problem.

4.3.3.2 Operational Behavior and Performance of Collectors

In actual operation at SSPS, the single axis tracking collectors have higher availability and need less maintenance, but they need higher insolation. The double axis tracking collectors have more frequent outages and require more maintenance efforts so their ability to operate at lower insolation levels is offset by the higher percentage of hours lost.

Analyzing daily values obtained from the «Monthly Data Reports» and concentrating on «normal» and «good» operational data, Schreitmuller was able to obtain good empirical fits for a total of 270 observations with at least one collector field in operation [II,3.2].

In Fig. 67 and 68 for the single and for the double axis tracking collectors, the thermal energy collected per square meter is shown as a function of the daily offered energy. The upper line denotes the thermal output related to

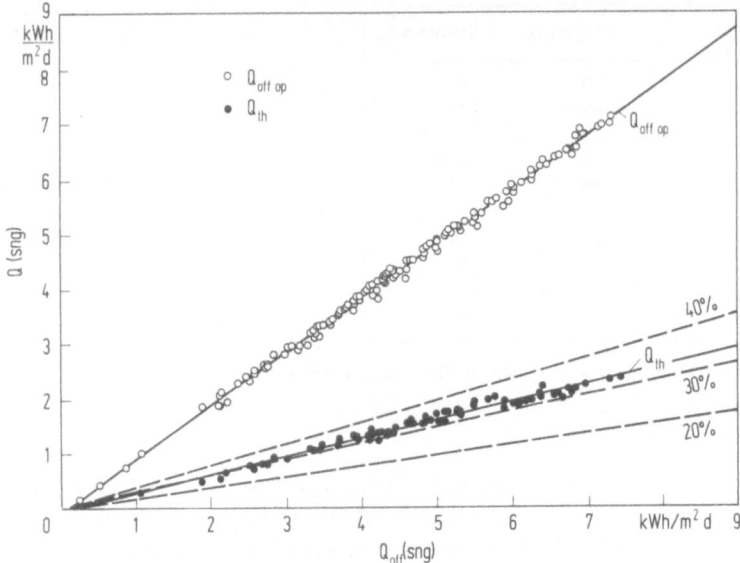

Fig. 67. Thermal energy output of single axis collectors

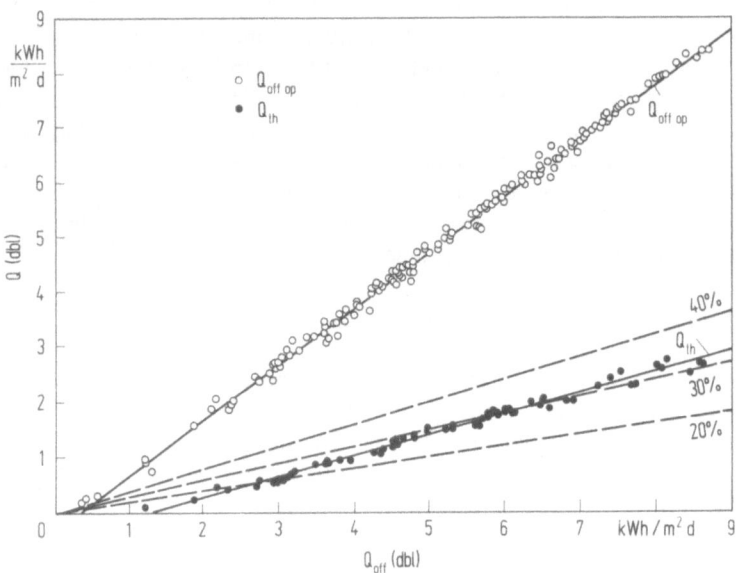

Fig. 68. Thermal energy output of double axis collectors

the energy offered during operation, i. e. without taking into account the energy available during outages and maintenance shutdowns.

The solid line in Fig. 67 fits the equation:
    Q(sng) = −0.12 + 0.349 * Q(off)
The solid line in Fig. 68 fits the equation:
    Q(dbl) = −0.421 + 0.362 * Q(off)

In these equations, 'Q" stands for thermal energy, "sng" or "dbl" for the single or double axis tracking collectors, and "off" for energy offered to the respective collectors (all Q in kWh/m /d).

Comparison shows that the double axis collectors have a considerably larger offset which is due to their much larger thermal capacity. This drawback is partially compensated by the higher slope in the second term of the equation. It implies a higher specific yield, as might be expected from their ability to track the sun in both coordinates. Schreitmuller also calculated the yearly output of thermal energy on the basis of the data available from the 270 operational days. As there were many sunny days when the DCS did not operate because of measurements, repairs or adaptations, he extrapolated the available information, making reasonable assumptions for a more reliable plant under typically commercial conditions of operation.

Type of tracking	gross total therm. kWh	per aperture area, kWh/m^2	per land area, kWh/m^2
single axis	1'032'022	385.95	122.9
double axis	986 838	367.13	50.7

Fig. 69. Comparison of yearly outputs of collector types [II,3.1]

4.3.4 Performance of the Thermal Storage System

This system did not present any particular operational problems, but as it was situated outdoors, periodical cleaning of components affected by rain and dust, as well as revisions of motor and changing pump packings did impose an extra load on routine maintenance jobs.

Andersson investigated the losses of the thermocline tank extensively [4.3] and reached the following conclusions: temperature measurements of the charged tank under stationary conditions indicate a loss rate of 17 kW during the daily cycle when the ambient temperature varies between 14 to 27 °C and the wind speed varies between 3 and 25 km/h. On this basis, tank losses amount to 4.1% per day; however, losses measured during normal operation amounted to values between 6.9% and 9.4%. Calculation errors as well as the difficulty of separating tank losses from piping losses during normal operation are suspected to be the cause of this discrepancy. The fact that the rate of the energy loss did not decrease appreciably during the 6 days of the stationary test indicates that the thermocline tank time constant is of the order of two weeks [CA] when piping losses are excluded and about one week when operational losses are included.

If the field needs to be stowed because of the weather or for operational purposes, there may not be enough hot oil to run the PCS. If the hot oil temperature were to fall below 275 °C at the time when the field can be put back into operation, it would be too cold to run the PCS and too hot to go to the collectors. Therefore it would be necessary to mix the oil in the tank by recirculating it through the fields in desteer and returning it to the top of the tank. Thereby energy is lost at a rate of 800 kW (energy loss rate: kWh/h) and the tank reaches a uniform temperature in 20 minutes.

The stability of the thermocline was checked by means of a thermocouple tree assembled over the entire height of the tank. Stability appears excellent; i. e., internal heat losses are quite small, which is certainly in accordance with the long time constant already mentioned.

The dual medium storage tank was found to be more difficult to operate. Testing of this tank, which was put into operation in 1984, was not yet completed at the time of writing this.

4.3.5 Power Conversion System

The power conversion system did not cause any unconventional outages and therefore did not need any unconventional maintenance efforts; however, there were some outages for repairs of a conventional nature due to injector blockages, broken seals and to a leaky weld in the hot oil piping.

In terms of performance, the PCS operated below the design conversion efficiency of 20% due to the scaling down of a 3 MW turbine to 0.6 MW, reaching a conversion efficiency of 17% from thermal to mechanical, as may be seen from Fig. 70.

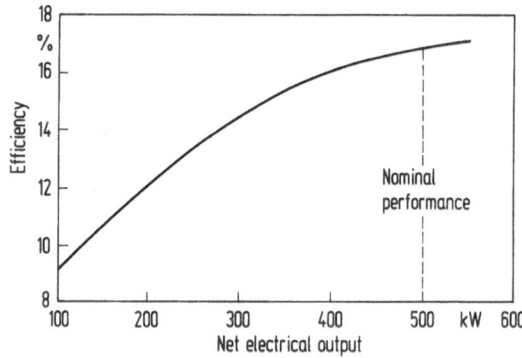

Fig. 70. Efficiency of steam turbine

The efficiency function shows that it is not reasonable to operate the PCS with a thermal input of less than 500 kW, as Schreitmuller [3.2] points out in his analysis of collector field performance. The PCS also needed a considerable amount of energy for daily warmup and pipe losses were high. The high thermal capacitance of the PCS is due to long distances between thermal components which in turn were responsible for large masses of water and steam. The design specification calling for fast load change capability is difficult to meet with such a design.

4.3.6 Simulations and Transient Behavior

After the plant became operational, a simulation code SESAM (Solar Energy System Analysis Model), was developed by Belgonucleaire under contract with the DFVLR (Andersson [II,3.5]). This code differs from similar computer programs: it was specifically developed for the SSPS DCS plant and is therefore efficient in modeling the actual hardware behavior. With the exception of the turboalternator, the whole DCS plant is modeled in detail. In interactive mode, the code is approximately 15 times faster than the real time operation of the plant. The code may also be run in batch mode

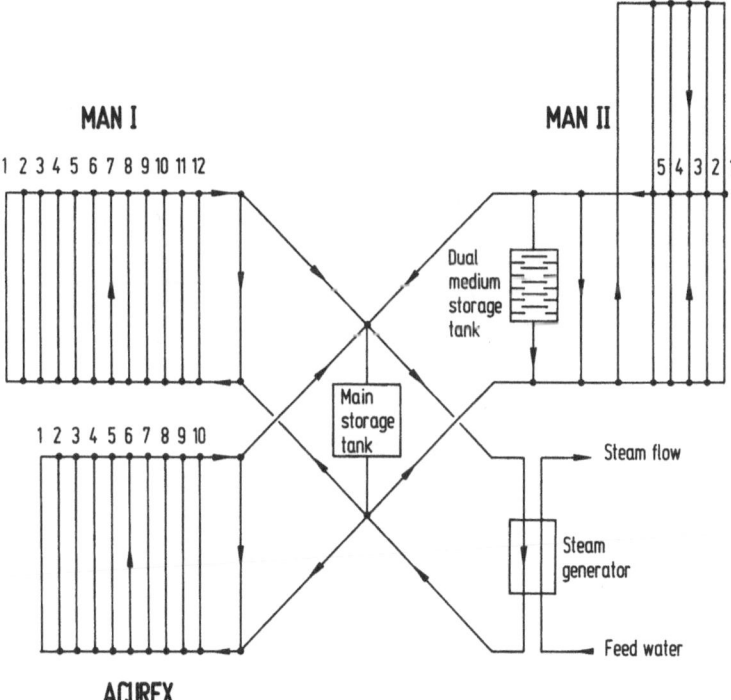

Fig. 71. The DCS plant as seen by the simulation code

whereby the whole day can be simulated in less than 10 minutes. Figure 71 shows the DCS plant as seen by the SESAM code.

Actual data recorded on a number of typical days were used in order to obtain realistic input parameters for the code. Data on thermal losses, thermal inertia and collector efficiencies were obtained to tune the code. The accuracy of the code was checked by comparing the results of the code to the measured behavior of the plant on one specified day as shown in Fig. 72. In this figure, DAS represents the curves obtained from the data acquisition system, and SESAM indicates the simulated curves. The quality of the simulation is so precise that, although checks were made on a number of days for both ACUREX and MAN collectors, only one set is reproduced here.

The simulation code was used to investigate the effects of a number of operational strategies and the following conclusions were drawn by Andersson [II,3.5 "Irradiance level for start up"]:

– What is the optimum irradiance level to start the fields? Andersson shows

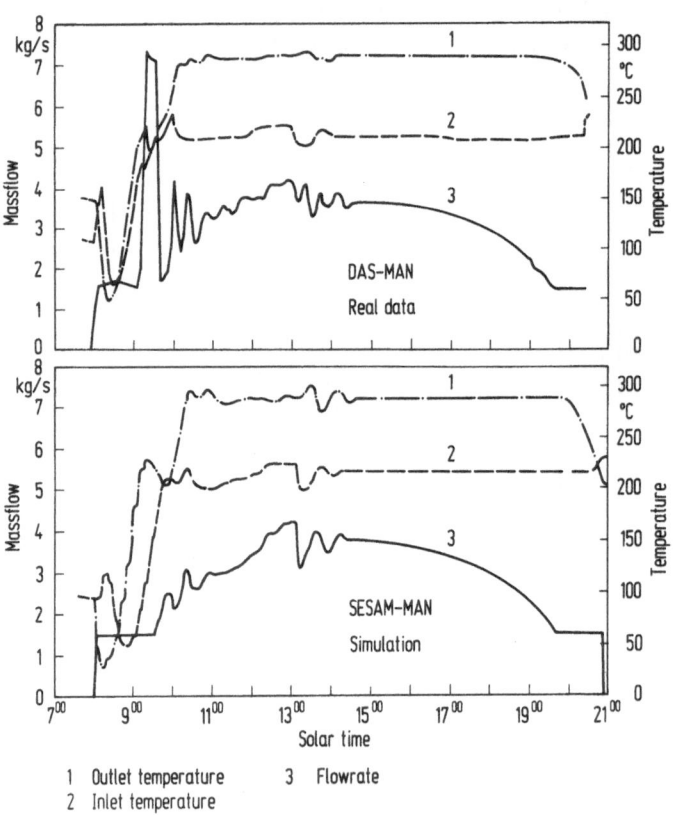

1 Outlet temperature   3 Flowrate
2 Inlet temperature

Fig. 72. Check of the accuracy of the SESAM simulation code

that startup is best when insolation reaches 100 W/m² in the aperture plane of the ACUREX collectors which corresponds to about 400 W/m² in the aperture plane of the double axis MAN collectors. With an insolation of 400 W/m² in the aperture plane the single axis collectors lose 2.4% and the double axis collectors about 1.2%.

– Lowering the operating temperature from 190 °C to 175 °C reduces thermal losses and thereby improves thermal efficiency by 4.7% for the MAN and 3.4% for the ACUREX collectors.

– Soiling affects reflectivity and lowers the optical efficiency of the collectors. Simulation shows that when reflectivity is reduced to 70%, the single axis collectors supply only 50% of the energy to the storage tank and the double axis collectors supply only 35%.

Transient effects investigated by simulation included modifying the bypass valve's operation strategy and the effect of reducing thermal inertia:

– In the existing strategy, the bypass valves are used to recirculate the oil in the field as long as the outlet temperature is below 240 °C. When the outlet

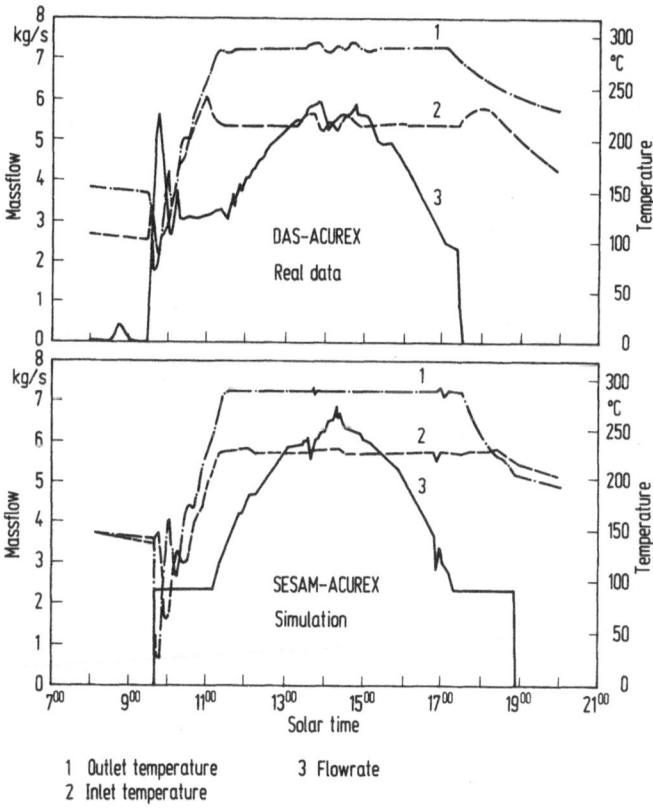

1 Outlet temperature   3 Flowrate
2 Inlet temperature

Fig. 73. Effect of change in bypass strategy

temperature increases, the valves start closing and at 275 °C all oil is sent to the storage. As the field inlet temperature drops when the PCS starts, causing undesirable field flow oscillations, a strategy was simulated whereby the bypass valves were controlled so as to keep the field inlet temperature constant. The beneficial effect of this change is seen in Fig. 73 in which the upper curves show actual conditions with flow rate oscillations and the lower curves show the probable improvement the simulated strategy change would make.

– Thermal inertia of the double axis tracking collectors makes long heating times necessary, especially on winter mornings. The large metal mass of the long pipes and the quantity of oil contained in them cause this high thermal capacitance. It was therefore desirable to simulate the effect of reducing the thermal capacitance of the MAN field. The results of this particular simulation can be seen in Fig. 74: reducing the thermal inertia by a factor of two allows the system to reach operating temperature almost an hour earlier.

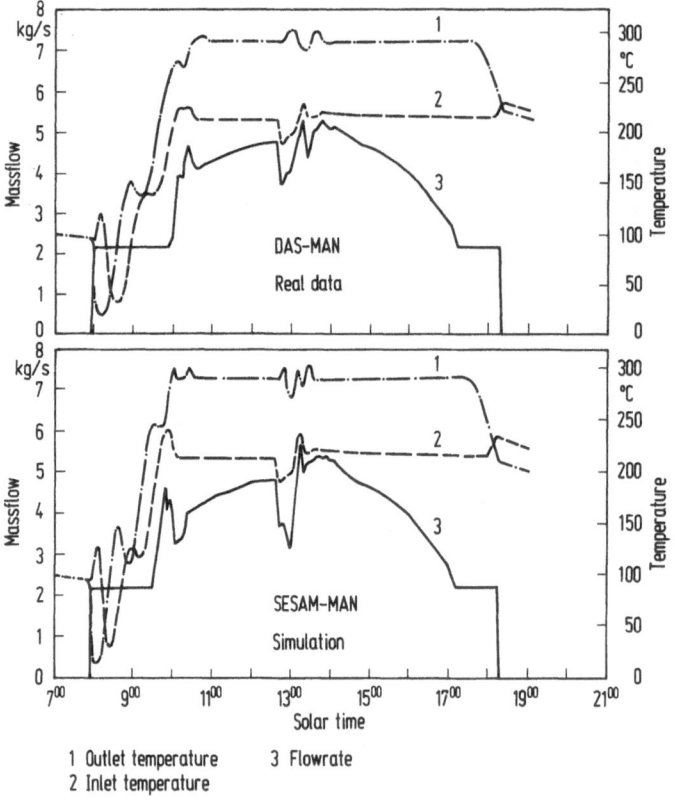

1 Outlet temperature   3 Flowrate
2 Inlet temperature

Fig. 74. Effect of reducing thermal inertia

For the effect of bypass strategy, see Fig. 73.

Note that the simulation is able to predict even reduced flow rate oscillations, as may be seen in the upper picture.

For the effect of reducing thermal inertia, see Fig. 74.

Here again, the simulation is quite capable to predict the "fine details" of the process quite accurately, as may be seen from the real and the simulated oscillations.

4.3.7 Production of Electricity

1. *First year of operation*

Wattiez reports that the DCS fields operated during 85% of the time when the insolation was above 300 W/m², collecting 35.2% of the energy offered to them [II,3.1]. During the month of June, when 290 hours of insolation above 300 W/m² were offered, the efficiency of the fields reached a maximum value of over 40%.

The PCS was in operation during 25% of the time when intensities above 300 W/m² were available. The efficiency of conversion from thermal energy at steam generator outlet to electric energy at alternator terminals was 18% averaged over this time period but two thirds of the produced electricity were needed to cover the internal consumption of the plant. Under these conditions, the total conversion efficiency from solar to electric averages out to 4.3% gross and 1.6% net deliverable to the grid. Because of the need for frequent functional adjustments during the first year, it would not be reasonable to evaluate the behavior of the plant by calculating the amount of energy produced.

2. *Second year of operation*

The performance attained during the second year is summarized in Fig. 75 (all energies in MWh).

	Hours of operation	Collected energy	Produced electricity gross	net	Conversion efficiency
ACUREX field	1744	850	–	–	30.6%
MAN field	2103	845	–	–	26.4%
Total	3847	1695			
to PCS		1280	234	40	18% gross

Fig. 75. Performance of DCS during second year

The 18% gross conversion efficiency includes the energy needed for preheating the PCS; about 11% of the thermal energy is needed for this purpose. About 500 kWhe are needed for both the DCS and the CRS plants to run control room equipment and DCS controls.

3. *Third year of operation*

In March of 1984 the size of the plant was increased by the addition of the third collector field with more MAN collectors. It was therefore necessary to go through a new period of adjustments and measurements so that the production of electricity was frequently interrupted during good solar days. The thermal energy increased by 23% and the internal consumption reached 74% so that the gross power output for the third year reached the same level as during the second year.

Energy losses of the enlarged DCS plant are shown in the next table. Calculations were based on 100% availability of solar energy between sunrise and sunset over the period of March to June 1984 as shown in Fig. 76.

Energy not caught by the collectors between sunrise and sunset:	26%
Energy not caught considering only irradiation above 300 W/m:	20%
Optical and thermal losses of fields while producing heat:	34.4%
Optical and thermal losses while tracking the sun:	63.3%
Heat losses in piping and storage:	2.5%
Preheating of collector fields:	18.1%
Preheating PCS:	1.4%
PCS operating losses:	14.6%
Internal consumption ("parasitic load"):	1.5%

Fig. 76. Main losses of enlarged DCS plant

Piping and storage amount to 11.8% of the thermal energy delivered by the field.

One should be careful not to consider the "small" percentage of parasitic load a negligible amount; the 1.5% refer to the solar energy offered taken as 100%. These losses actually represent about two thirds of the gross production of electricity.

The PCS preheating loss must also be considered because it amounts to a significant fraction of the thermal energy delivered to the PCS.

In spite of all the interruptions experienced during the third year, the plant still was able to produce 153.4 MWhe gross energy through the year.

Schreitmuller [II,3.2] has used data gathered in 1983 to obtain what may be considered a reasonable extrapolation of the yearly production of the DCS. Assuming a typically commercial plant whose operation would be interrupted only for routine maintenance tasks, Schreitmuller obtains an idealized gross output of 253.2 MWh with a gross conversion efficiency from solar to electric of 3%. Under these conditions the net electric output would be 138.6 MWh and the net efficiency 1.63%. For the period of October 1982 to September 1983 Wattiez reported a gross efficiency of 4% averaged over the entire year and 5.3% when averaged over the net production period [II,3.1].

When looking at these efficiencies one has to bear in mind that the DCS had to supply a significant part of its production to run all subsystems and experimental facilities associated with the project; a difficult task for a small plant. In addition, it took too much heat to bring the steam turbine drive to operating temperature, and its thermal to mechanical conversion efficiency was small because of its modest power rating and its low operating temperature. The latter could be increased even for a DCS by using larger trough collectors; such a plant is presently in construction near Barstow, California. Also, in a larger DCS plant the internal consumption would amount to a smaller percentage of the gross output power thus improving the net conversion efficiency.

5 General Aspects

5.1 Comparison Between CRS and DCS Power Plants

One of the original goals of the SSPS project was to compare the CRS and the DCS plants. Because of the system limitations encountered during their construction and operation, meaningful comparisons can be made in some selected areas of interests, but only as long as operational parameters, such as availability or differences in design points, are taken into proper consideration.

For example, during the experimental phase of the project, between November 1981 and August 1984, the CRS plant was only in full operation for 101 days. It is a novel and complex system; it was necessary to take many measurements requiring defocussing the heliostats or shutting off particular subsystems; there were some unanticipated "exotic" failures, such as the very frequent breakdowns of the steam motor (which had been designed for less demanding tasks) and a lengthy series of repairs involving inadequate design and poor workmanship in one of the storage tanks, causing problems which required much time to find solutions; so it was impossible to operate the plant during as long an uninterrupted period as the DCS could be operated. To compare the net electric energy produced by each of the power plants would therefore not reveal the true reliability nor the true average efficiency of the CRS. For the same reason, a comparison of investment costs or of the net cost of the produced electrical energy would not yield a useful answer. An attempt to extrapolate the potential yearly production of the CRS is equally doomed, as it would be necessary to assume widely differing outage statistics for different models without having any degree of confidence in the models assumed.

Nevertheless, some useful comparisons can be made and the results do permit to draw certain safe conclusions regarding further development efforts, or the construction of operational systems or the determination of the best application for the two concepts.

Before making any statements about the different behavior of the CRS and the DCS it is appropriate to compare their major parameters, such as done in Fig. 77. For simplicity, the three collector fields of the DCS have been taken as one; the differences between the fields are treated later in the comparison.

Parameter	CRS	DCS
Mirror area, m²	3657	7602
Land usage factor	0.2	0.27
Geometric concentration factor		
for cavity absorber area	215	40
for ASR absorber area	466	
Mean operating temperature	400 °C	250 °C

Fig. 77. Comparison of main design parameters [EFN, p. 117]

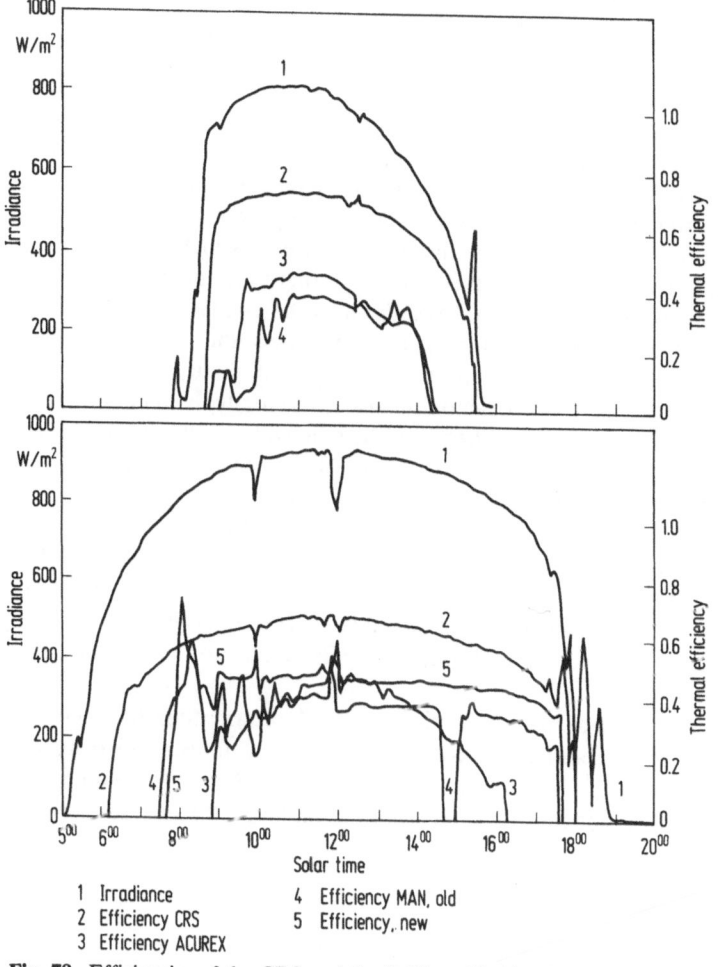

1 Irradiance   4 Efficiency MAN, old
2 Efficiency CRS   5 Efficiency, new
3 Efficiency ACUREX

Fig. 78. Efficiencies of the CRS and the DCS on 12. 12. 1983 and on 6. 7. 1984 [EFN]

Figure 78 shows a comparison of the efficiencies of the CRS and the two original DCS fields for a medium winter day and for a medium summer day.

The thermal efficiency of the CRS is superior to that of the DCS under all conditions normally encountered; as Sandgren and Andersson point out, [EFN] it produces twice as much thermal energy per reflector area as the ACUREX field. Inspite of the higher temperature, the CRS has an efficiency of approximately 60% compared with about 37% for the DCS fields.

Due to its higher optical concentration the CRS starts collecting solar energy much earlier than than the DCS in spite of its greater thermal inertia. The difference is approximately one hour in winter and two hours in summer.

At the time of writing there is not sufficient operating experience to determine the extent to which less availability offsets the advantages of higher energy collecting efficiency, nor the exergetic advantage of the CRS plant over the DCS for equal power ratings.

The DCS is easier to operate and to automate. It requires less personnel and can be considered an already marketable system for some selected applications, such as for pumping in arid zones, for the production of process heat, and especially for coproduction [9] of thermal and electrical energy.

The CRS concept gives promise of improved conversion efficiency and of lower production costs than the DCS at larger power ratings. However, it is not yet possible to determine the size or power at which the crossover of production costs can be expected, although there are indications that this may occur somewhere between 10 MWe and 30 MWe peak power rating [3].

As far as environmental factors are concerned, there appear to be no significant differences between the two types of plant; the CRS has a somewhat bigger land usage factor but also (potentially) a higher efficiency. Also, it appears that the land between the heliostats can be put more easily to agrarian use than the land between the DCS collectors. Neither plant emits any significant amount of noise or noxious product.

In spite of the major difference in the nature of the power conversion systems adopted for the two SSPS plants, the thermal inertia of both generator drives was surprisingly similar; the CRS steam motor as well as the DCS steam turbine both absorbed too much thermal energy before reaching full power conditions. In addition, both PCS conversion efficiencies were plainly too poor.

In terms of termal storage, neither of the adopted systems had a clearcut advantage. Sodium and oil are both non-optimal substances for storing heat in a compact space. They both require rather large surface to volume ratios for the storage tanks causing extensive thermal losses through surfaces and supports.

5.2 Economic Assessment

5.2.1 Introduction

The intent of the IEA-SSPS project was to explore the technological possibilities of exploiting solar power at small power levels. From the beginning, investigations were limited to gaining insight into the necessary technology only, and no provisions were made to explore the cost aspects, which would have required the construction of a variety of different sized plants with different power ratings. Also, as about one dozen solar plants were being constructed simultaneously in a number of different places, the SSPS EC was able to gain significant insight into cost/size relationships by closely watching corresponding developments in other countries; in fact, many members of the SSPS team were simultaneously involved with other projects of different magnitude. Thanks to such connections, in April 1983 the EC was able to establish a working group for the economic assessment of solar power plants compared with conventional means of generating electricity. The group was chaired by Prof. G. Faninger of ASSA (Austria) and was formed by W. Bucher (DFVLR Köln Porz, GFR), M. Geyer and H. Klaiss (DFVLR Stuttgart, GFR) and J. P. Thornton (SERI Denver U. S. A.). The following summary was compiled from the report of this working group [3], from considerations expressed by P. Kesselring of EIR (Switzerland) in one of his papers [6], from the results of the feasibility studies of solar power plants in the alps [5], and from the leading work of K. W. Battleson [2].

5.2.2 Economic Analysis

As the working group points out, conclusions on economic aspects cannot be drawn decisively from the experience gained so far with solar thermal power plants, because their experimental nature has not permitted sufficiently long, undisturbed operation. The 10 MW plant in Barstow is the only one which has been in commercial operation long enough to warrant a reliable judgement on plant economics and efficiencies. Therefore, the economic aspects of 10, 30 and 100 MWe peak power rated CRS plants, and of a 10 MWe rated DCS plant, have been evaluated in accordance with certain specific data.

Heat generation, cogeneration, hybrid cycles and combined cycles are all considered important but have not been included in this particular study.

The basic methods used were: the annuity method, and also a number of life cycle cost methods, generally used for conventional power plant evaluation because they form a basis for sensitivity analysis. The quantitative results presented in the study were obtained with the life cycle cost methods.

Subsidies, tax incentives and so on, were not taken into consideration as they vary from place to place and from time to time.

The results presented in this study were obtained from the following assumptions:

- yearly beam radiation: 2500 kWh/m^2,
- price base: 1983,
- depreciation for the 10 MW CRS: 15 years,
- depreciation for the other plants: 30 years,
- discount rate: 9%.

The results of the study are summarized as follows:

- Different methods of cost analysis lead to different electricity production costs. For example, the real levelized cost method results in 0.33 DM/kWh for the 100 MW CRS plant while the nominal levelized cost method gives 0.52 DM/kWH. The relative ranking of the different alternatives is not affected: the nominal levelized cost always leads to higher values than the real levelized cost method.
- For a projected 100 MW CRS the costs amount to 0.33 DM/kWh while for the installed 10 MW CRS plant the costs are 1.85 DM/kWh.
- Using the real levelized cost method in both cases, the production costs come to the same order of magnitude for solar thermal and for conventional power plants; 0.12 to 0.20 DM/kWh.
- As can be expected, specific investment costs decrease considerably with upscaling plant size; however, the costs listed for the 10 MW CRS were actually paid for an already built plant, whereas the costs for the 100 MW plant are projections into the near future.
- The sensitivity analysis shows the following factors to be of significant influence: investment cost, life expectancy, discount rate and yearly energy production.
- In all cases, investment costs amount to over 90% of the total costs while maintenance and repair costs are of comparatively low influence. Specific investment costs vary between 10 000 and 20 000 DM/kW.
- The lions share of the investment costs goes to the heliostat field.
- No particular type of receiver or of heat transport medium has shown any significant cost advantage. The choice therefore has to be made on technical and operational requirements.
- A cost optimized storage could not be found in spite of the analysis of storage capacities between 70 minutes and over 10 hours.
- As only single point data are available for the DCS, it is not possible at this time to obtain a cost ranking or a cost crossover between DCS and CRS power plants but it is assumed that for larger plants, the CRS will be superior.
- The specific costs of actual solar power plants are 3 to 10 times higher than those of conventional plants; however, the costs of conventional electri-

city are steadily increasing while the costs of solar electricity show a continuous decrease thanks to further research and development efforts.
– The study concludes that there is a realistic chance for solar power plants in the future and that the process would be accelerated if fuel prices and government support were increased.

Figures 79 through 82 are reproduced from the Faninger report because of the importance of the underlying assumptions and the results of the study.

		10 MW	30 MW	100 MW
Output	MWe	10	30	100
Full power hours	h	2185	2533	3815
Yearly el. output	kWh/a	27.3*10	75.6*10	419.7*10
No. of heliostats		1818	1877	15500
Mirror area	m^2	39.5	95	57
Height of tower	m^2	77	125	200
Cooling medium		water/steam	sodium	fused salts
Storage time in minutes		170	70	510
Internal consumption	%	25	10	11
Personnel		(16–30)	27	67
Type of field		round	north	two north fields

Fig. 79. Input data of the CRS plants [3]

		DCS system		CRS systems			
Prices,	DM/kWh	10 MW	10 MW	10 MW	10 MW	30 MW	100 MW
Life expectancy	a	15	30	15	30	30	30
Real present value		1.32	1.02	1.85	1.37	0.50	0.33
Annuity method		1.63	1.39	2.34	1.96	0.70	0.47
Nom. present value (levelized costs)		1.78	1.56	2.42	2.09	0.77	0.51

Fig. 80. Specific production costs of electricity [3]

Although Fig. 82 gives personnel costs for a minimum of 16 people for a 100 MWp CRS plant, according to more recent information, 25 people would be a more realistic figure. However, with 16 people, specific operating and maintenance costs were calculated between 100 to 400 DM/kWe which would have no significant influence on total costs. This conclusion is not strongly affected by a small increase in personnel for plant operation. Operational improvements such as automation and the normal increase in the component reliability with operational experience will certainly help reduce the necessity for human intervention.

It is interesting to compare the different estimated costs as calculated in studies made on both sides of the atlantic. Already in 1981 Battleson used

Investment data	10 MW			30 MW			100 MW		
	10⁶ DM	DM/kW	%	10⁶ DM	DM/kW	%	10⁶ DM	DM/kW	%
Heliostat field	100.5	10 500	34	94.0	3 140	30	354	3 540	31
Spec. costs (DM/m)	140.0	–	–	530	–	–	400	–	–
Receiver	38.0	3 800	13		no data		105	1 050	9
Thermal cycle	53.4	5 340	18		no data		61	610	5
Tower	23.3	2 330	8		no data		66	660	6
Storage	29.0	2 900	10		no data		140	1 400	12
Site preparation	7.0	700	2		no data		21	210	2
Electrical equipment	11.6	1 160	4		no data		58	580	5
Miscellaneous	33.2	3 320	11		no data		320	3 200	28
Total inv. costs	296.0	29 600	100	94.0	3 160	30	320		

Fig. 81. Investment costs of CRS plants [3]

Operating and Maintenance	10 MW		30 MW		100 MW	
	10⁶ DM	DM/kW	10⁶ DM	DM/kW	10⁶ DM	DM/kW
Personnel	2.5	250	3.2	107	6.5	65
Repair and Maintenance	1.0	100	2.0	67	3.9	39
Machinery costs (variable)	0.3	30	0.65	21.7	3.5	35
Total costs	3.8	380	5.85	195.7	13.9	139

Fig. 82. Yearly operation and maintenance costs of the CRS plants [3]

data from 28 different solar power projects to make a cost analysis similar to the one previously described [2]. Based on his assumptions, Battleson obtained the levelized busbar electricity costs as a function of the capacity factor; for a heliostat cost of $ 200/m^2 and a capacity factor of 0.3, levelized busbar electricity costs are 16.5 Cents/kWhe. The corresponding figure for specific heliostat costs of $ 97/m^2 is 7.5 Cents/kWhe. Increasing the capacity factor to 0.6 levels the costs of electricity at a value which is lower by about 20% to 25% lower (The "capacity factor" in a way measures the availability of the plant, without taking forced outages and planned maintenance shutdowns into account).

Using data from the SOTEL [4] and the METAROZ [5] studies, Kesselring [6] quotes specific electricity costs between 0.57 and 0.90 SFr/kWh for CRS type plants projected for a variety of sites in the Swiss Alps. However, some of the sites considered were technically not optimum because, in Switzerland, inexpensive land with good solar statistics, and without legal protection from construction for scenic reasons, is difficult to find.

The same paper quotes cost estimates for electricity produced by solar dish stirling driven induction generators and by photovoltaic power plants. Depending on the "learning factor", costs of 0.38 $/kWh decreasing with time to 0.21 $/kWh are quoted for the dish sterling and initial costs of 1.12 $/kWh decreasing to 0.21 $/kWh are predicted by the proponents of the dish sterling approach.

The effect of the so called learning curve has been used in most studies; it relates the costs of the heliostats (and other subsystems) to the quantity produced with time, According to Battleson, initial heliostat cost may be as high as 5000 $/m^2, but could be reduced to below 100 $/m^2 when production approaches 250 000 a year. Battleson and Faninger used the same estimates in their cost analysis, but in order to reach this low a cost figure, significant advances in technology and in manufacturing techniques may also be required.

5.3 Environmental Impact

In order to thoroughly examine the advantages of a new technology in comparison with an existing technology, all environmental effects, both positive and negative or damaging, have to be taken into consideration, as well as other economical, social and political factors.

Land usage is certainly one of the significant impacts made by solar power plants. Where land is of premium value, the large area required is a major consideration and a source of concern to ecologists and economists as well. According to Battleson [2] a 10 MW solar power plant with three hours of storage would occupy some 2.5 km^2 in a favorable solar site in the southwestern US. Assuming a yearly availability of 1 500 hrs (peak power equivalent, i. e. more than 1 500 hrs actual operating time, but not always at full power), yearly production would amount to 15 000 MWhe. An artificial lake in the mountains with a hydroelectric plant capable of providing this amount yearly may actually require more than 2.5 km² if the geopotential difference or if the yearly precipitation were not high. Such plants do exist in large numbers and their "lakes" have a severe impact on local ecology since rivers are reduced to insignificant brooks and the large variations in the water level prevent the shores from being used for agrarian or recreational purposes.

In southwestern Spain, a number of hydroelectric plants which were built in the Twenties have lost up to 70% of their production capacity due to longterm climate changes and also to the fact that improvements in irrigation techniques have led to better use of water so that the tributaries to the reservoirs are running dry. These plants are also operated by the CSE and one is tempted to suggest that they be converted to thermal solar electric, preferably using the site formerly covered by the lake.

Contrary to popular opinion, the choice of sites for solar power plants is not restricted to deserts and cheap real estate: in some mountanous areas there are sites with very good solar statistics and a very clear atmosphere, where the snow actually helps by washing the heliostats. At other sites the topography alone lends itself either to the placement of a solar receiver or to the construction of a DCS plant. Many of these sites would be otherwise quite useless, so the impact on the local environment would be tolerable.

Another negative impact of a solar thermal plant is the consumption of water for cooling and cleaning purposes. With everything else equal, a solar thermal plant will use the same amount of water as a fossil or nuclear plant with the same yearly energy output; but because the capacity factor of the solar plant is larger, the cooling load is at times much greater than that of a fossil or nuclear plant which runs steadily at constant load. However, at the power levels presently considered, this is probably not a major problem on many sites. Battleson computes a yearly requirement of 2 million m³ of water for cooling tower makeup [2]. The local impact of the steam from the cooling tower has to be considered in populated areas as has already been found necessary with conventional plants. Obviously the placement of the cooling tower should also take into consideration the shadowing of the heliostats or collectors of its own plant by the emerging steam as it might be blown by possible wind directions.

There are several residues from power plants, whose effects on the environment should be considered, although some of them (such as sewage, trash and remains of lubricants) are common to many non solar installations. The runoff of deionized water from washing the mirrors does not in itself damage the ground water or the soil, but it does enrich the original water with sodium, potassium, calcium and magnesium salts. Modern reverse osmosis plants are usually designed in accordance with local regulations to keep the impact on the environment within legally accepted limits.

Paving of the surface around the heliostats or collectors creates a different type of drainage problem. Natural vegetation is eliminated and precipitation has to be channeled in order to prevent erosion damages to the site and to adjacent areas. However, paving is not always necessary on certain rocky grounds typical of mountanous sites. On the site of the Shenandoah solar power plant near Atlanta, GA (USA), the climate allows to preserve the natural ground cover between the solar collectors thus reducing the runoff problem and additionally giving the advantage of a very significant reduction of the local atmospheric pollution by dust.

In all projects of this sort, plants and wildlife are affected negatively during construction and afterwards. No general statement can be made about the desirability preventing such effects or rehabitating the flora and fauna after construction. In many cases, wild animals are capable of adapting to the new conditions and learn to live with them. At Solar One it is a rare event when a bird attempts to fly through the focus of the CRS. Since solar power plants do not emit noxious fumes or loud noises, it is to be expected that many animals can live near them.

As 15 to 20% of the insolation received at a solar site are converted into another form of energy, there will be some local shift in climate. Of these 15 to 20% most is converted into heat and released through the cooling tower and a smaller fraction is "exported" from the site in the form of electric energy which is ultimately converted into waste heat at the consumer's site. But unless very high power levels are reached, this conversion of insolation to electricity will only affect the temperature distribution of the air masses near the ground and not the local climate.

Particular care must be taken to avoid accidental or careless spills of harmful substances such as mineral oils, silicones, sodium, solvents etc. which are harmful to humans, animals and plants and can severely damage the ground water, sometimes for prolonged periods of time. In this respect, the same rules apply as for any other technical installation in which such substances are routinely handled. Most countries have adequate safety codes which can be adapted to the operational requirements of a solar power plant.

Last but not least, the impact on the environment caused by manufacturing the components required for the construction and operation of a solar power plant must also be considered. The factories which produce the steel, aluminium, glass, sodium, high temperature oil etc. all emit some form of harmful residue and need energy supplied by other plants. Although it is too early to determine the magnitude of this impact, it is an important consideration. The energy payoff time and the total energy balance of solar power plants should be calculated as soon as sufficiently reliable data become available.

Consideration of the global effects of the extensive use of solar energy yields an interesting picture:

The total amount of solar power impinging on the earth is approximately 15 000 times greater than all the power produced by human technology. As mentioned above, on a local level the thermal impact of a solar power plant is quite minor; therefore on a global scale it is not expected to have any significant effect either. So if one considers the effects on the climate caused by the contribution of carbon dioxide to the atmosphere from burning fossil fuels, and if it is assumed that a considerable part of the energy obtained from fossil fuel combustion could be substituted by "clean" heat such as, for example, solar power, then the environmental impact of solar power utilization should be positive. This holds true for the production of electrical power as well as for other uses of solar thermal energy.

6 Lessons and Guidelines for the Future

6.1 Site Selection Criteria

The SSPS project has taught an important lesson regarding the selection of appropriate sites. Conventional solar statistics issued by airport authorities and tourist offices can be quite misleading.

An accurate knowledge of the number of hours per year of clean beam radiation is not sufficient. Factors such as circumsolar radiation, beam irradiance as a function of energy density thresholds (such as shown in Fig. 4, section 2.2) and soiling characteristics of the sites under consideration are of paramount importance. Consider for example a possible site in a mountainous region with almost no other known meteorological conditions than clear sunshine or lots of good, clean snow: the beam radiation statistics would perhaps not be better than those of an arid subtropical zone, but the percentage of time during which the beam would reach intensities above, say 700 w/m^2, could exceed that of some "classical" desert site; the sunshape might be excellent most of the time due to the lack of atmospheric pollution, and the snow would clean the little dust accumulated on the heliostats. Such observations are presently being made in the Alps by Durisch [5] and others; the results obtained so far are quite encouraging.

Guidelines for the future

The selection of an appropriate site is of prime importance. Direct measurements of daily insolation intensity and sunshape on all sites under serious consideration should be taken for at least one year. At the time of writing it is already possible to obtain excellent synoptic data on cloudiness and surface temperature (as an indicator of insolation intensity) from weather satellites. Once appropriate radiation statistics are available, curves of the type presented in Fig. 4 (section 2.2) can be used to determine the number of heliostats or collectors necessary to obtain the desired solar multiple. Note that Fig. 4 shows that the average monthly insolation can be approximated by a straight line for beam radiation levels above 700 W/m^2 thereby greatly simplifying the development of mathematical models for the evaluation of sites and for the determination of the required solar multiple.

Many other physical characteristics are of importance for the selection of sites such as:

- frequency and duration of cloud passagees,
- wind, storm and lightning statistics,
- effects of precipitations on the ground surface,
- runoff from nearby mountains,
- suitability of the ground for foundations and underground piping.

Further selection criteria include:

- accessibility of the site for transports and personnel,
- availability of technical services and conventional spare parts,
- availability of suitable personnel,
- possible impact on local industries and farming efforts,
- political and social factors of local character such as community attitudes and relationship with labor unions.

According to Battleson [2] the solar multiple is defined as "The ratio of thermal power absorbed in the receiver fluid and delivered to the base of the tower at design point to the peak thermal power required by the turbine generator (or other end use)." The "solar multiple" therefore depends on the "system design point", which in turn depends on the calendar day selected for the basic design specifications; however, as has been stated earlier, knowledge of the radiation received during the design point day is not sufficient to calculate an adequate solar multiple.

Guidelines

An expression is needed to describe the site specific insolation pattern during the entire year, because the solar multiple as presently defined is inadequate. Clearly, either the definition of the term "solar multiple" should be improved, or another term should be invented to describe "excess mirror area" necessary to obtain a given performance, such as the yearly output of net energy. The term should account for insolation quality at the site under consideration and also permit optimization of the plants' thermal storage capacity.

According to Kesselring [6] careful consideration should be given to the tradeoff between increasing the number of heliostats and increasing the thermal storage capacity of the plant. Adding only heliostats leads to storage overflow while adding only storage capacity is useless if storage is already available most of the time. In order to achieve lower energy costs, it may well be necessary to increase heliostat area together with storage capacity while keeping the power rating of the PCS the same. Obviously, the present definition of the solar multiple is not adequate to determine the optimum ratio between heliostat and storage costs.

The available literature gives little or no information on the possibility of optimizing thermal storage in regard to production costs or operational factors. Simulation programs should therefore be developed to enable a very quick modeling of the entire system so that a large number of technical options can be studied in regard to their cost effectiveness before they are built. The emphasis for this particular application should not to be put on high accuracy, which is neither attainable nor necessary, but on computational speed.

6.2 Determination of Solar Multiple

Experience gained from the SSPS plants indicates very clearly the danger of underrating the solar multiple: beam radiation statistics show that with a marginal solar multiple, reasonable power production is only possible for a small part of the year.

The principal interaction between thermal storage capacity and solar multiple is explained by Battleson [2] in Fig. 83.

One has to bear in mind that thermal storage systems add quite significantly to investment costs and to the operational complexity of solar power plants. According to Battleson, a plant capable of operating 24 hours with a peak insolation of 950 W/m would need a solar multiple of over 2.5, which

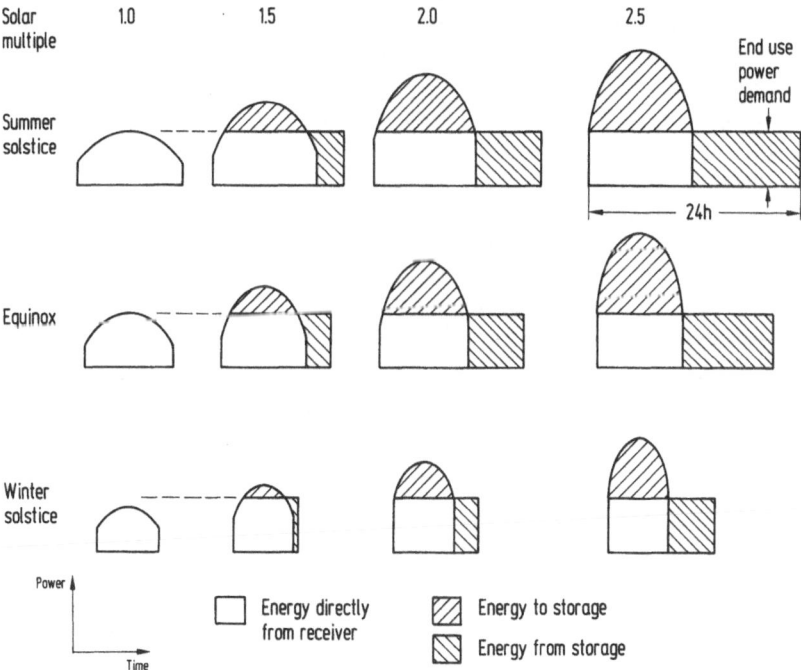

Fig. 83. Relationship between solar multiple and storage capacity

means it would not only be quite expensive but it would also waste a lot of solar energy on many days. Under these conditions there are conflicting requirements between land usage, heliostat costs and thermal storage costs on one side, and the gain in plant energy output on the other side.

6.3 Solar Specific Components

The SSPS project provided a number of unique opportunities to gain experience with solar specific components for a variety of reasons:

- Two different solar plant concepts, the CRS and the DCS, were built and operated side by side.
- Sodium technology, today only applied to nuclear power development was utilized for solar power generation for the first time and at relatively high power densities. Sodium proved its worth as a heat transfer fluid in the receiver area and in the steam generator. Its thermal conductivity, its relatively high boiling point and its excellent hydrodynamic and wetting properties enable the construction of primary fluid circuits with relatively light and thin walled tubing made from conventional materials.
- Two entirely different types of sodium receivers were developed and operated under practically identical conditions.
- A steam motor was used to drive an alternator at a relatively high power level and with specifications bordering on the limit of feasibility for piston driven steam engines.
- Design specifications, as well as data acquisition, simulation and control problems typical of a small system made it necessary to develop very specific software.
- In both systems marginality of the thermal energy available for conversion forced more attention on the characteristics of the solar specific components than if surplus thermal energy had been available.
- The very high availability of the heliostats and the collectors proves that these typically solar specific components have reached commercial status.

Although some very important lessons were learned regarding the solar use of non-solar specific components (see next chapter), their experimental nature caused very frequent shutdowns. Sometimes the shutdowns were due to failures, but more frequently they were imposed by the need to measure one subsystem or another, or by changes in the pertinent hardware, such as the switch from the cavity receiver to the billboard receiver. The intention was not to maximise production during the test phase but to provide a basis for future efforts; in that sense, the "outages" were important experiences. The operational phase was subject to the experimental purpose of the project, so that to some extent, operational experience with the solar specific components was not conclusive.

Guidelines

In order to prevent extensive damage to the HC and HFC due to lightning stirkes, the electronics should be "NEMP – hardened", as is customary with modern military and civilian high reliability systems.

To obtain a complete picture of the long term operational characteristics of the solar specific components, the continuation of prototype projects such as SSPS well into an operational production phase lasting several years is desirable. Such efforts could probably be maintained at reasonably inexpensive levels compared with the cost of scrapping or mothballing. Relatively simple economic studies could compare the costs of simply running the plant, and gathering operational data with costs of scrapping it.

6.4 Solar Application of Non Solar Components

SSPS experience regarding this can be summarized pointedly:

"When conventional components are applied to solar technology, there is no guarantee that they will behave conventionally"

The main reasons for this can be summarized with an acceptable degree of simplification:

– The daily solar cycle is two orders of magnitude shorter than the heating cycles of fossil fired or nuclear heated power plants.
– Solar radiation, which is thermal input, is highly stochastic.

This means that, conventional components are exposed to a large number of unconventionally short periods of thermal input, resulting seals, valves, pumps and control systems being brutally cycled through temperature ranges difficult to simulate in advance.

Another consequence of this "driving function" is that day to day shutdown times often exceed the power-on duration, unless a large solar multiple coupled with long term thermal storage is available. Under these conditions, the efficiency of the PCS also varies between low and normal values so that the average conversion efficiency falls short of the design efficiency for the entire power conversion from solar to electric.

Guidelines

To qualify for solar specific applications, non solar specific components must be subjected to design philosophies and quality control procedures of the type applied to military components and known as "mil-specs". The SSPS experience is uniquely qualified and could supply inputs for the creation of "sol-specs".

Fast running computer programs, efficiently simulating the different control strategies, would be desirable to determine the interactions between solar radiation conditions, hardware design and control options. Prolonged operational experience would give an opportunity to update the simulation software as soon as the learning progress allowed or when there were changes in the hardware, in weather patterns or in the operating conditions for other reasons.

6.5 Storage Subsystems Optimization

The sodium system is only attractive for thermal storage because of its simplicity. As other substances, such as molten salts or mixtures of liquids and solids exhibit higher specific heat and lower costs, it is evident that sodium is not ideal as a thermal storage medium.

According to Bucher, Geyer et. al., [3], no optimum size could be found for a thermal storage system on a purely economic basis. The picture would change if the main function of storage were to be merely the avoidance of interruptions due to cloud passages: a reserve of about 10 to 20 minutes should be sufficient, as longer cloud passages are usually associated with cloudy days when the available solar energy would be insufficient for reasonable power production anyway. Large storage systems have been built [I,4.3] to keep plants running over night (the connection between storage capacity and solar multiple is discussed in G.2), but in actual experience, these systems are best employed for covering overnight losses and trace heating requirements. Considering the large investment cost associated with the construction of such systems, their cost effectiveness appears doubtful; however, this is a purely intuitive judgement, and obviously, more precise data on the subject are needed.

Guidelines

Numerical codes calculating ideal capacities for storage systems must include not only the major subsystems, but also a simulation of the entire power plant. The running times of conventional digital codes are too long to allow contemplation of a large variety of design options from which the appropriate ones can then be selected as inputs for the economic optimization of the thermal storage system; more time-efficient methods should be developed.

A possible approach would be to simulate the thermal dynamics of the whole system using closed form equations, as known in the analysis of transients in electrical engineering, thus avoiding long run times for purely numerical integrations. The differential equations generally used for transient analysis are of the type

$$S(t) = Ax + Bx' + Cx''$$

In this equation $S(t)$ represents the solar input as a driving function and x the temperature at the steam generator input as a function of time t. The constants *A, B,* and *C* depend from the subsystems thermal capacities and from their thermal losses, parameters easily obtained from available data. After successful development of a simulation in closed form, the model should be extended to include the characteristics of the PCS. This is done best by using an empirical fit to the measured conversion efficiency; it would use the simplest possible closed form function rather than a highly accurate polynomial, thus favoring computational speed over extreme accuracy.

6.6 Thermal Inertia and Losses

In theory, thermal inertia is acceptable in that it enables a solar power plant to cover "solar outages"; yet at the SSPS project experience indicated that the low conversion efficiency of morning insolation by both the CRS and DCS was actually due to thermal inertia. In actual fact, however, the waste of morning insolation is not caused by thermal capacitance but by the night cooling of the thermal systems and the consequent need to replenish the lost heat. The term "thermal inertia" should therefore be defined more precisely to include its individual effect on each component or subsystem.

The need to keep sodium in a molten state requires energy consumption which is undesirable in energy limited systems such as solar power plants; either very carefully designed thermal insulation is needed, or the sodium-potassium eutectic alloy, NaK, has to be used. This is more expensive, more corrosive and more difficult to handle at room temperature than pure sodium; in addition, the need for trace heating would not be entirely eliminated since NaK has to be protected against freezing, also the heat transfer characteristics of NaK are somewhat poorer than those of sodium. Considering these aspects, it can be assumed that the choice of pure sodium as a heat transfer medium for SSPS was correct.

Guidelines

Future designs of solar thermal power plants should stress compactness of layout in order to minimize thermal losses through pipes. A compact design could include integration of thermal storage in the CRS receiver, integration of the PCS in the central tower, etc. For the DCS the total piping length could be minimized by locating the PCS in the center of the collector fields.

Subsystem support components which carry or store thermal energy should be designed so that metallic structures do not act as thermal bridges. Some techniques for reducing thermal losses have been developed in cryogenic technology: they include suspension by wires or glass fibers instead of massive metallic stands, the use of ceramics for parts held in compression, etc.

Components actually using thermal energy, such as steam generators and turbines should be designed so that the housing and the turbine wheels have a minimal heat absorbing capacity. The fact that this "unconventional" design approach would lead to more expensive heat exchangers and turbines can be justified when considering that the cost of the turbine is not a major part of the cost of the plant, and that the improvement of the overall efficiency of the plant could be worth the increased cost of some components.

6.7 Operating Strategies

Solar power plants with high solar multiples and large thermal storage capacities will be less sensitive to operating strategies than plants with marginal thermal input such as SSPS. Being particularly sensitive to operating strategies, statements can be made which apply in general manner to both plants:

- Overnight thermal losses in pipes, receivers and collectors in conjunction with the relatively high thermal capacitances of these exposed components are responsible for the long time required for preheating in the morning.
- The low solar multiples of both plants prevent full power operation while simultaneously charging the thermal storage tanks.
- Both PCS are inefficient, in particular at power levels below design specifications.

Guidelines

If energy limited systems are to be considered, careful optimization of operating strategy during the preheating phase is a must; not only the beam radiation quality but also the actual thermal condition of the storage, the collectors and the piping must be accounted for by the control system. The PCS should not be run as long as storage is not sufficient for full power operation. Load following is only reasonable with a full storage tank, unless large conversion losses are accepted in penalty. Fast real time computing with good simulation software is imperative for these control systems.

6.8 Internal Consumption

Internal consumption, often called "parasitic power", is a burden for both plants. To some extent it is due to their experimental nature, but with an output rating of 500 kWe, it would be difficult to reduce the internal power requirements of any power plant to a reasonable proportion as long as there were fluids to be pumped, valves to be operated and mirrors to be moved. Even so, a considerable learning process took place at SSPS as evidenced from the detailed accounts of operational experiences.

Guidelines

It is not possible to give general guidelines on reducing internal consumption because it depends very strongly on the characteristics of the hardware in use. The reader is therefore referred to section 6.9 which discusses possibilities for improvements.

6.9 Recommendations for Future Efforts

There is a distinction between possible improvements of present plants and the desirability of developing new technologies for the utilization of solar power in purely thermal applications.

The first goal leads to a broader fulfillment of the original SSPS project objectives, letting it grow from an experimental to a truly operational power plant and thereby obtaining a long range assessment of the potential of power plants of this size. Long range technical, economic and ecological aspects of plants with about one megawatt of electric output could then be compared with larger systems, and scaling laws, useful for the synthesis of national and international power programs, could be derived. Such scaling laws have already been developed by Weingart, [12] Battleson [2], Bucher [3] and others, but their scope and range in size does not cover plants of this type, so decision makers might be tempted to rely on intuitive judgements.

The history of technological innovation is full of examples in which intuitive judgement and preconceived notions or oversimplified assessments have led to wrong decisions. A case in point is the development of commercial jet transport: after the US Air Force paid for the development of the KC 135 aerial tanker, a number of proponents fought to have it converted into a civilian passenger aircraft. Several aircraft manufacturers studied the economics of this possible new generation of transports and came to the conclusion that only a limited number of such expensive aircraft would be sold and that the commercial risks of the venture were too high. The economics of the propeller airplane were well known and, at the time, it was correct to assume that the jet transport would cost more to build and to operate. When it was finally decided to go ahead with the development of the passenger aircraft, the very manufacturer of the KC 135 underestimated the market of the "707" by more than one order of magnitude! It was only after the Boeing Company had firm orders for some 200 aircraft that competing companies entered the battlefield, some of them too late, as history has shown. Today propeller aircraft have lost their dominance in passenger transportation in spite of the fact that for some applications, propeller driven aircraft can be more economic.

The second goal, to develop new applications for solar power technology, is at least equally important. It is reasonable to expect that the knowledge of the typically solar parts of SSPS can be used for developing new means of exploiting solar radiation for purposes other than electricity production with an "already paid for" starting condition. The potential of new technologies can not be proved before the hardware is actually built and tested. No amount of systems analysis would have proved the practical usefulness of the laser when its physics were just discovered, but once laser technology was at hand, its usefulness became obvious. Luckily, no amount of systems analysis can predict the difficulties that will beset new developments; if this were so, nobody would find the courage to engage in shaping our future.

6.9.1 Suggestions for Improvements of the CRS

Productivity of the present CRS plant could be increased significantly by taking a number of measures, some of which would be very cost effective:

1. To improve the solar multiple (which is clearly insufficient), the number of heliostats should be increased: now that meaningful beam radiation statistics have been obtained, the number of heliostats needed to supply the subsystems at their present size with a minimum level of beam radiation will be easy to determine. A larger solar multiple might improve plant production statistics very significantly even without increasing the capacity of the present thermal storage. Simultaneously, a study should be made to determine the extent to which increased production due to an enlarged storage capacity would offset the increased price for the system.
2. Consideration should be given to repositioning some of the heliostats, as it has been shown that a more optimal placement is possible [8.2].
3. To prevent corrosion, a way of supporting the mirror modules without encasing them should be adopted. A system is presently being tested on site which appears to be more impervious to corrosion than the actual one.
4. To prevent extensive damage to the heliostat control systems, the HCs and the HFCs should be "EMP-hardened".
5. A heliostat washer truck of the type used in Solar One should be used to improve the average mirror's reflectivity.
6. Both receivers have proved their strengths and their weaknesses and both concepts can be significantly improved; based on experience gained at the SSPS project it would appear possible to build a new receiver without engaging in an expensive development program which has already taken place. The new receiver would not only have improved overall efficiency but would also simplify operations such as sodium filling and startup.

Cost effective design improvements are therefore possible with both receivers:

- The cavity receiver could be reduced in size and the spacing between the absorber tubes could be eliminated; the exposed surface would be reduced from $62\,m^2$ to $24\,m^2$ thereby reducing losses considerably. As the two bottom tubes make no significant contribution they could be eliminated and the tube bends could be insulated in front and behind [I,5.4].
- The ASR sodium circuit could be changed to enable a conventional filling operation thereby avoiding complex and risky procedures. A change in circuitry has been investigated by Maffezzoni [I,7.1] and appears to present no particular flow control problems.
- For both receivers, tilting down would further reduce convection losses.

7. In an energy limited system such as SSPS all intentional cooling should be avoided whenever possible. For example, a sodium pump construction which does not require cooling would waste less thermal energy. Solutions should be investigated which allow pumping at temperatures above $500\,°C$, such as driving the pump through a long ceramic shaft or perhaps even using direct pumping of the sodium by means of magnetohydrodynamic principles (MHD).

8. Trace heating and startup losses could be entirely eliminated if a small oil (or natural gas) fired boiler were added: the benefit to solar power exploitation would be considerable, because PCS productivity on good solar days would be more than doubled and would be at least possible on many days when the present system needs too much preheating time to reach the production phase. The CRS would still produce all its energy from solar power and "waste" a little fossil fuel instead of a lot of its own electricity. In future plants, the thermal storage system could be reduced, or might even not be necessary if a fossil fuel trace heating system were chosen.

   This particular low-cost improvement appears reasonable even if the number of heliostats could not be increased. The cost/benefit ratio of this type of trace heating concept should be carefully evaluated to establish its worth.

9. A larger fossil fired boiler, capable of supplying all the thermal power for full PCS operation, would still not require major changes in the solar specific components and would make the SSPS a "hybrid" plant. The plant would produce electricity from fossil fuels, but the savings in fuel consumption, in the order of 20% to 30%, would be an appreciable improvement impossible to attain without the use of solar power.

10. Since the hot and cold storage tanks lose a significant amount of heat through their supports, it has been suggested that these be changed; the problem of heat losses through supports is well known in space technology and in the construction of energy-conserving window frames.

11. Since solar power plants require a relatively large investment effort, maximum use of their operating time and minimum use of manpower

should be made. In consequence, extensive automation is a must and personnel shift policies must include weekend operation of the plant.

6.9.2 Possible Improvements of the DCS

DCS technology shows evidence of having reached an almost commercial level. However, experience gained at SSPS allows some conclusions to be drawn and some guidelines for future applications of distributed collector systems have been derived from it:

1. As the thermal storage and transport system is responsible for a significant part of the losses, the PCS should be installed in the center of the fields to minimize pipe length. The cost effectiveness of the dual medium storage tank is low and it is very heavy. If dual medium storage is desirable, tanks filled with gravel and sand (such as used in Barstow) should be used.
2. Nearly two thirds of the electrical energy is used by the plant itself. The requirement to increase automation conflicts to some extent with the requirement to reduce internal consumption; therefore, a fresh look at the layout of the electrical systems is needed.
3. An isolated economic analysis comparing costs and yield of the two different types of collectors (single and double axis tracking) would help select the type of collector appropriate for each particular application.
4. Washing is needed 12 to 15 times a year and is cumbersome because it is difficult to wash the modules with an automated device. Washing has to be done manually and requires the use of a platform. Means of automating the washing operation should be studied.

6.9.3 The Author's Opinion

As an outsider, I have formed an opinion which is largely my own, however, it must be recognized that this opinion may to some degree have been influenced subconsciously by the discussions with the people involved in the project. Although total objectivity may seem desirable, the reader is reminded that a totally "unopinionated" document would not only be quite boring but would also not stimulate his outlook on the future possibilities of the described technology. I therefore hope, that whatever opinions show through, this book will contribute towards awakening an interest in the potential of solar thermal technology, as it did with me as a writer; being an electrical engineer with experience in photovoltaics and with no objections to the reasonable use of nuclear power, I had not paid much attention to solar thermal power utilization until I became involved in writing this book.

Was SSPS a cost effective effort? During the general solar thermal overoptimism of the early 70ies, SSPS was originally intended to consist of two different types of prototype plants capable of routine operation, thereby allowing valid comparisons between CRS and DCS. Due to a number of unanticipated difficulties and in part also to monetary restrictions, the two plants never reached the prototype status; from a utility point of view, routine power production was not achieved. Nevertheless, recognition of this fact led to a more or less tacit decision to concentrate on the problems of solar thermal power utilization and on finding solutions to solar specific problems. I am convinced that, at a cost of less than 30 million US $ the ratio of "knowledge gained" to "money spent" was indeed favorable. Oddly enough, this was due precisely to the subcritical power rating of the two plants which accentuated most solar specific problems, forcing the engineers to place the emphasis of their work on research and development. A higher power rating would have yielded plants which would have been closer to production prototypes (as has been pointed out in sections 6.2 and 6.5); but the ratio of knowledge gained to money spent would have been less favorable.

From experience as described in the foregoing chapters, it is evident that the expected crossover of specific electricity costs between CRS and DCS was not found, but that DCS technology is much closer to commercial reality than CRS technology. On the other hand, since DCS operating temperature is lower than that used in CRS plants, the Carnot efficiency of DCS is limited to lower values. It is therefore reasonable to assume that the use of DCS plants as a source of process heat may in many cases be economically more attractive than their use for the production of electricity. CRS plants have more promise for the production of electrical power and can be expected to yield a higher thermodynamic conversion efficiency.

Simulation methods account in large measure for the success of the SSPS project: Numerical simulations were used extensively in the design, the testing and the determination of the characteristics of the plants' components; as a matter of fact, all subsystems designed with the aid of simulation programs performed close to expectations. It is therefore fair to say that numerical simulation must be considered an absolute necessity in the development of similar projects, right from their planning phase: now that the necessary experience has been obtained and the empirical data can be used as parameters for simulations, it is easy to use simulation methods to determine the adequacy of sites.

Very significant progress was also made in the development of software for economic assessments of solar power plants. In a way, it can also be considered "simulation" software, as it opens the way to a connection with systems optimization simulations, as described in section 6.5. This connection is an absolute necessity for broadening the spectrum of options open for future cost optimization efforts.

Because of the pioneering nature of the project, instrumentation methods often had to be developed ad hoc. On many occasions in the field the ITET was forced to deal with solar specific situations and parameters for which there were no off-the-shelf methods or instruments. Much of this methodology and instrumentation will be required in future efforts for the utilization of solar power, be it in the construction of other solar power plants or in the development of future technologies. From similar experience in other fields of endeavor, one can only wish that the developers of this "exotic" instrumentation will find the time and means to produce exhaustive descriptions of their instruments and methods and the software associated with their use. To mention just one example of the accomplishments of the researchers and developers involved in the SSPS project: methods were developed to measure the energy density up to 2.5 MW/m inside receivers, without physically contacting the surfaces subjected to high temperatures. Such methods can be expected to find application in many other fields of high temperature technology.

It is fair to say that the SSPS project has left us with a research facility built at a reasonable cost, easily accessible from western Europe and not overly hard to reach from across the Atlantic. The facility could easily be adapted to a number of solar oriented research and development efforts. It is being run by a staff which has built up a lot of solar specific experience and is accustomed to cooperating with international teams. The host country is interested in supporting high technology work and has a stable government. The site is adequate for solar research precisely because it is not quite optimally suited: factors which affect solar power utilization negatively, such as haze, dust, winds and lightning, force the engineers to make realistic assumptions and to work under realistic conditons.

All questions concerning the quality of the international management within the SSPS project were answered without hesitation in a positive manner and all people interviewed considered the international cooperation as one of the encouraging experiences of SSPS. On the basis of these experiences, a continuation of many selected R+D efforts at the SSPS site would be extremely desirable even if some planning time were to elapse before the necessary decision process came to the point of implementation. It should also be borne in mind that some countries are using the same site, just a few hundred meters from the SSPS, to investigate the feasibility of producing 1 MWe by means of gas cooled receivers [8]; this fact adds to the attractivity of the site because it increases the locally available knowhow and widens the local facilities. To start a new project elsewhere, with different conditions, and a team without this rich experience, would certainly be more costly and would hardly speed up success.

Some thoughts on the potential usefulness of solar power plants in general are in order at this point:

In the face of hydroelectric, fossil and nuclear "competition" solar power plants have yet to find their place in the spectrum of options for producing power; however, one particular application comes to mind when visiting Solar One in Barstow: the majority of its power is produced at times when it is most needed in the city of Barstow for cooling purposes. It stands to reason that the cost of bringing electric power with feeder lines from far away to desert cities would be reduced if solar power were locally available. Obviously, an economic tradeoff should be studied for cases where long and expensive high voltage lines are needed to supply cities with high cooling loads.

In some semi-arid areas of southern Spain, hydroelectric power plants which were built during the beginning of our century are no longer capable of delivering their originally rated power because the tributaries to their lakes are increasingly being used for agricultural purposes. In cases where the electricity production is of local importance (and also to prolong the utilization of the existing grid) it would be reasonable to investigate the possibility of building solar power plants near the lake thus substituting the loss of hydroelectric potential with solar power. Since precipitation and sunshine are opposed to each other, power would come predominately from the water or the sun, depending on the weather conditions. Another striking possibility of this type of "hybrid" plant is the capability for long term storage of solar electric power by letting the hydro plant use solar electricity to pump water which has already run through the turbines back into the lake; this particular scheme to store excess electrical energy has been in use succesfully in Switzerland for half a century, and it exhibits overall storage efficiencies of between 60% and 70%. One particular power plant located roughly 30 km north of Sevilla (owned and run by the CSE) near the village of El Ronquillo, appears particularly suitable because the weather satellite pictures indicate much less cloudiness for this area than for Almeria. Similar conditions exist at another plant called "El Pintado", some 60 km north of Sevilla, which was built during the late fourties. In both cases, solar power plants with peak ratings of the order of 50 to 100 MW would greatly upgrade the usefulness of the existing installations. A scheme is also conceivable whereby a solar "energy storage plant" built at the lower end of the hydro conduit would use its steam turbine to run the pumps directly thereby reducing the inefficiencies of the multiple conversions involved in the previously described system. With a full lake, such a plant could also power the hydroelectric plant directly, thereby avoiding pipeline losses.

Finding "niches" for solar power utilization, be it for the production of solar electricity or otherwise, could become a very rewarding effort. Modern tools for numerical analysis and simulation make the task feasible, and without such studies, it cannot be stated with certainty that such niches are uninteresting.

6.9.4 Future Research and Development Efforts

The motivation for developing further uses of solar energy in the future is hampered by the present situation: the technologically well developed countries are still experiencing strong economic growth, in spite of a yearly reduction of about 2.5% of the world consumption of mineral oil. For this reason, according to H. Steeg [10], politicians tend to overrate the importance of the situation and repress future problems in spite of the fact that anticipated total energy consumption will increase by about 23% by the year 2000. As the development of new sources of oil (shales and sands) is very costly, and as the environmental impact of fossil combustion may become a threat in the future, it is reasonable (if not necessary) to assume that it is not too early to start developing new uses for solar energy.

In central and northern European countries, the large scale use of concentrating collectors for high temperature applications of solar power does not appear very promising due to rather poor insolation statistics; exceptions my be found on some alpine sites. However, the utilization of global radiation by means of area collectors for heating water or air, and the use of photovoltaics, appears more promising. Experiences in Germany and in Switzerland indicate that solar architecture may be the best way to make use of these resources; also, systems designed to dry grass for feed stock, to heat swimming pools, to supplement the production of warm water in housing projects and other similar low temperature applications, have proved economically successful. From the thermodynamic point of view, however, using a source with a temperature of some thousand degrees such as the sun, to heat washing water to a mere 50°C seems like an appaling misuse of an exergetically valuable source.

This last thought warrants paying special attention to the possibility of using solar energy for high temperature processes, thereby obviating the need for exergetically more valuable electricity, or for environmentally more damaging sources of heat. According to Kesselring [6], the cost of a thermal kWh is already comparable to the cost a kWh produced by oil combustion; should it be possible to build a cost effective receiver for endothermal processes, a "solar reactor", as chemists might call this type of processing vessel, might be an extremely useful device for developing a new, environmentally benign industry in sunny countries. This type of technology would open a number of new possibilities, such as:

- manufacturing synthetic fuels and chemicals, thereby making solar energy storable and transportable,
- pyrolyzing "valuable" wastes such as automobile tires and other rubber products yielding hydrocarbons which could be used as fuels or as base products for the synthesis of plastics,
- pyrolyzing assorted household wastes with similar results as above,
- research into high intensity photochemistry, a field on which little or nothing has been written,
- research in high intensity photophysics such as the photovoltaics of gallium arsenide or investigating new laser schemes,
- substituting combustion processes which are harmful to the environment,
- making process heat available in areas which are devoid of conventional fuels.

One particularly interesting example for the two last cases would be the reduction of cinnabar to mercury by using solar heat, as Kesselring [6] has suggested. It is customary to reduce the mineral by heating it with oil burners or mineral coal and the process liberates sulfur dioxide, mercury sulfide and other noxious reaction products. Could mercury be obtained by using clean solar heat, the only resulting byproduct would be essentially pure oxygen! In Almaden (Spain) some 40% of the world supply of mercury is produced and it would not be surprising to find sites with good insolation statistics not too far away from there.

If it becomes possible to absorb concentrated solar radiation directly onto particles or droplets with efficiency, it might be possible to store energy by endothermal chemical processes thereby obviating the need for thermal storage, one of the major problems of solar energy technology. This scheme would be of particular interest, if processes could be identified which did not require exact matching of the quantity of educts in the reaction space with the momentary radiation input; in other words, processes which are rather insensitive to the stochastic nature of the solar energy entering the reaction chamber. Such "heat input insensitive" technologies do not only exist, some of them are very old, such as the reduction of iron ores and the calcining of limestone.

The principles of direct absorption and conversion are relatively unknown, and even if their impact on a future "energy problem" should turn out to be minor, an appropriate R+D effort would increase the knowledge of the problems and advantages of this type of "radiation chemistry". The side effects of such investigations should also be considered: better and larger quartz windows might have to be developed, high power light pipes developed for this type of research could become useful for the transmission of energy in special cases, simulation and optimization software would probably become useful in other fields etc. The interchange of information between this field and others, such as space technology and ecology, would constitute an additional benefit which is difficult to assess without actually becoming involved in this activity.

One might counter that this type of sophisticated technology has little chance in the climate typical of central Europe or most of the USA; yet Winter, Nitsch, Klaisz and Voigt [11] point out that, although this direct utilization of sophisticated technology may only have limited chances in well developed countries, many of these countries need to export power gear and complete power plants in order to survive economically. This is particularly true of Switzerland which has no mineral resources. The development of a "solar industrial capability" in countries with poor insolation is therefore not as absurd as it may seem at first; it may even be more economical to build solar facilities where insolation is less but where capital costs are low because of low interest rates and low inflation.

The energy problem is of an international nature and the benefits of having international teams doing research in the relatively new field of solar technology are obvious. Research in the utilization of solar energy in no way endangers the success of other types of energy production as solar power cannot fully substitute many of the classical schemes, such as nuclear power production. On the contrary, the operation of the present power plants is necessary not only to support our civilization, but also to support all those developments which are not yet capable of self supporting productivity; this has always been the case throughout the history of technology. Whether solar technology may or may not, some day, compete with fossil fuel technology in general, or just in some particular applications, cannot be determined at this time. What is certain is that it will open new doors and that we need to know what lies behind these doors.

Glossaries

Introduction

These glossaries have been prepared as a contribution to the book being prepared by Professor Federico Casal on the International Energy Agency's Small Solar Power Systems (SSPS) Project. The Glossary on solar thermal energy conversion has been compiled from a number of sources, including reports prepared by the author[1,2] and others[3] together with solar thermal documents[4,5,6,7] from Sandia National Laboratories and DFVLR.

The Glossary on solar radiation was developed using a variety of sources including the classic works by Boes[8] and Robinson[9]. Especially useful is the document[10] by the IEA dealing with meteorological measurements and data handling for solar energy conversion applications. The recent IEA document[11] on meteorological and environmental conditions at the Almeria site also provided useful definitions directly relevant to the SSPS project. A concise financial glossary has also been provided, based in part on a solar electric power plant financing guidebook[12] developed by the author and his colleagues at Polydyne, Inc. in San Mateo, CA.

Solar Thermal Energy Conversion

Absorber
The portion of the receiver that absorbs radiant energy.

Absorptance
The ratio of the radiant flux absorbed in a body of material to the radiant flux incident upon it. Also defined as "the fraction of the transmitted light incident on the receiver surface that is absorbed".

Absorption
The process in which incident radiant energy is retained by a substance. A further process always results from absorption. The irreversible conversion of the absorbed radiation into some other form of energy within and according to the nature of the absorbing radiation.

Absorptive Coating
>An absorber coating which improves the absorptance of the absorber to radiant energy.

Albedo
>The ratio of radiation reflected by a surface to that incident upon it.

Angle of Incidence
>The angle between the central ray incident on a surface and the normal to the surface at the point of incidence.

Annual Average Solar Efficiency
>The ratio of the annual solar energy delivered to a thermal process divided by the product of the annual direct normal insolation times the heliostat reflective area.

Attenuation Loss
>Loss of solar power by absorption and scattering due to atmospheric conditions. Losses are caused by scattering by air molecules, by selective absorption by certain molecules (e.g. water vapor), and by absorption and scattering by aerosols.

Availability (Operating)
>The percent of time the unit was available for service, whether operated or not. It is equal to available hours divided by the total hours in the period under consideration, expressed as a percentage.

Availability (Solar)
>Percentage of daylight hours for which the solar radiation level exceeds the threshold for plant operation (ca. $300 \, W/m^2$ for the DCS plant).

Base Load
>The minimum electric utility load in a given period of time. Also refers to a power plant designed for continuous operation at its rated capacity.

Beam Alignment
>The adjustment of individual mirror facets of a heliostat to place their images in the desired relationships to one another.

Beam Characterization Subsystem
>A system for the rapid and automatic measurement and characterization of flux delivered by any single heliostat.

Beam Quality Error
>One standard deviation (RMS) of the difference between the isoflux contour that contains 90% of the heliostat total power of a perfect heliostat (*i.e.*, no canting or mirror waviness errors) and the desired focal length, and the actual isoflux contour containing 90% of the heliostat actual total power. This error is in the heliostat reflected ray coordinate system. (Units: milliradians).

Blocking

The interception of part of the reflected sunlight from one heliostat by the backside of a second heliostat.

Bottoming Cycle

The lower temperature cycle in any energy conversion system where two (or more) separate cycles are used in cascade fashion (exhaust of one feeds input to another). See *Topping Cycle*.

Brayton Cycle

The thermodynamic cycle upon which combustion turbines are based.

Buffer Storage

The use of some form of thermal energy storage for decoupling the transients associated with the energy source from the end use process of energy, typically less than one-half hour of storage.

CRS

Central Receiver System

Capacity

The maximum power output rating of a generating unit or plant.

Capacity Credit

The amount of generating capacity displaced by a solar power plant, expressed in MWe or as a fraction of the nominal solar plant output.

Capacity Factor

Energy production in a given time interval (generally annually) divided by the energy that would have to have been generated if the end use were to be operated at its full capacity for the same time interval.

Cavity Receiver

A solar energy receiver in the form of a cavity where the solar radiation enters through one or more openings (apertures) and is absorbed on the internal heat exchanger surfaces.

Central Receiver Power System

(also known as *Central Receiver Power Plant, Central Receiver Plant,* and *Central Receiver System*) – see *Solar Thermal Central Receiver Power Station.*

Closed-Loop System

In reference to thermal energy transport and storage systems, one in which no part is vented to the atmosphere.

Cloud Cover

That portion of the sky cover which is attributed to clouds, usually measured by a trained observer in eighths or tenths of the sky covered.

Collector Efficiency

The ratio of the energy collection rate of a solar collector to the radiant power intercepted by it under steady state conditions.

Collector Subsystem

An array of individually controlled heliostats, including the wiring and controls, that redirects the available solar radiation onto a receiver.

Concentration Ratio

The ratio of reflected radiant power impinging on a surface divided by the radiant power incident upon the reflecting surface.

Cosine Loss

The reduction of projected heliostat area visible to the sun caused by the tilt of the heliostat, proportional to the cosine of the angle of inclination of the normal of the heliostat surface to the sun's rays.

Cost Effective

The system design alternative with the lowest cost/performance ratio.

Cost/Performance Ratio

A measure used in comparing system design alternatives wherein both cost and system performance are accounted.

Cost/Value Ratio

A measure used in evaluating how the cost of a system over its lifetime compares with the value of its product (*e.g.*, energy).

DCS

Distributed Collector System. A solar thermal electric system comprised of line focus parabolic trough collectors or paraboloidal concentrator units. The 500 kWe DCS plant at Almeria consists of two types of line focus parabolic trough collectors, organized in three separate fields. Thermal energy is collected by the troughs using a high temperature oil that serves for thermal transport, storage, and exchange, the latter for steam generation.

Design Point

The insolation conditions for which system performance is specified. For the SSPS central receiver plant, the design point is an insolation value (direct beam normal) of 920 W/m^2, corresponding to expected conditions at noon on March 21. Under these conditions, the system is designed to deliver its rated capacity of 500 kWe.

Diurnal

Daily; having a daily cycle.

Downcomer

The pipe carrying the hot heat transport fluid down the tower of a central receiver system.

Dual-Medium Storage Tank (DMST)

The second state of the DCS system is a dual-medium storage system in which thermal energy is stored by means of thermal oil and cast iron slabs.

Emissivity

The ratio of the radiant energy emitted by a surface to that emitted by a blackbody at the same temperature.

End Use

The final use of the thermal output of a solar central receiver plant, e.g., in a turbine to generate electricity, or in an industrial process.

External Receiver

A solar energy receiver where the solar radiation is absorbed on the external surface.

Fixed Charge Rate

The amount of revenue per dollar of capital expense that must be collected annually to pay for the fixed charges associated with plant ownership, e.g., return on equity, interest payment on debt, depreciation, income taxes, property taxes, insurance, repayment of initial investment, etc. It may also include operations and maintenance expenses expressed as a fraction of the capital cost.

Heat Tracing

Auxiliary piping heating system to prevent freezing of liquid (*e.g.* sodium) within the pipes.

Heat Transport Fluid

The fluid used for transporting or transferring thermal energy from one area to another within the system. See *Receiver Fluid, Working Fluid.*

Heliostat

A combination of mirrors, support structure, drive mechanism, and mounting foundation that tracks in two axes of motion to continuously reflect the sun's rays onto a fixed receiver.

Heliostat Field Efficiency

The ratio of the solar radiant power into the receiver cavity aperture or onto an external receiver under specified reference conditions, to the product of the insolation and total heliostat field reflective area.

Heliostat Packing Density (Ground Cover Ratio)

The ratio of total reflective surface area to the total land area used by the heliostats (collector subsystem).

Heliostat Specific Weight

Weight of the heliostat, excluding the pedestal and foundation, divided by the reflective area. (Units: kg/m^2)

Hours of Storage

The number of hours a solar plant can produce power at a stated output level, normally at full rated system load, when operating exclusively from an initially fully-charged thermal energy storage unit.

Hot/Cold Tank Storage

A thermal energy storage system utilizing separate tanks for the charged (hot) and uncharged (cold) storage medium.

Hybrid

A power plant using both solar energy and a nonsolar energy source (*e.g.* natural gas, oil).

Intercept Factor

The fraction of direct or reflected rays incident on the receiver aperture whose trajectories reach the absorber.

Maximum Receiver Thermal Power Rating

The maximum thermal power at the base of the tower of a CRS that the receiver will deliver sometime during the year.

Nameplate Rating

The full-load continuous rating of a power plant under specified conditions as designated by the manufacturer. (Units: MWe or KWe)

Optical Efficiency

For a solar thermal electric power plant, this is generally considered as the product of four parameters: reflectance, intercept factor, transmittance, and absorptance.

Parasitic Power, Parasitic Energy

The parasitic power is the power required at any time to operate the power plant (*e.g.* pumps, motors, computers, lighting, air conditioning, etc.). The parasitic energy is the energy consumed by such uses for a specified period. The *net power* produced by a solar thermal plant is the *gross power* generated less the parasitic power losses, and similarly for *net energy* production.

Peak Load

The maximum electrical load in a given time interval.

Penetration (Solar)

The solar power plant capacity as a percentage of the utility grid capacity.

Plant Availability

The percentage of time a plant is able to generate power if so required. Since there is an insolation threshold below which the plant cannot generate net power, the *availability factor* is defined as the fraction of total daylight hours when the solar irradiance exceeds the threshold value (*e.g.* 300 W/m^2 for the DCS system).

Pointing Error Per Axis

The standard deviation (RMS), for each axis, of the difference between the desired aimpoint and the beam centroid location. This error is in the heliostat reflected ray coordinate system. (Units: milliradians)

Power Tower

A term used in the popular press to describe solar thermal central receiver power systems.

Process Heat

Heat used for agricultural, chemical, or industrial operations.

Rankine Cycle

The thermodynamic cycle upon which water/steam turbines are based.

Receiver

That element of a solar central receiver system to which solar radiation is directed by the heliostats and where it is absorbed and converted to thermal energy.

Receiver Efficiency

The ratio of the thermal power absorbed by the receiver working fluid and delivered to the base of the tower to the solar radiant power into the receiver under the reference conditions.

Receiver Fluid

The working fluid that is circulated through the receiver to absorb the solar radiation as thermal energy, also called *heat transport fluid.*

Reflectance

The ratio of the reflected radiant flux to the incident radiant flux. For a solar thermal electric system, this is the *specular* or mirror reflectance of the collector surface.

Repowering

The retrofitting of existing fossil-fueled power plants or process heat plants with solar energy collection systems in order to displace a portion or all of the fossil fuel normally used.

Riser

The pipe carrying the cold heat transport fluid up the tower of a central receiver system.

Shadowing (or Shading)

The shading of the reflecting surface of one heliostat from the sun's rays by another heliostat.

Solar Multiple

The ratio of thermal power absorbed in the reciver fluid and delivered to the base of the tower at the system design point to the peak thermal power required by the turbine-generator (or other end use).

Solar Only

The operation of a hybrid power plant (or repowered plant) on the solar energy subsystem output alone. See *Stand-Alone.*

Solar Thermal Central Receiver Power System

(Also known as *Solar Thermal Central Receiver Power Plant. Solar*

Central Receiver Plant, and *Solar Central Receiver System*) A solar power system which concentrates the available solar energy, using an array of computer controlled heliostats to redirect the sun's rays to a tower-mounted receiver. The energy absorbed at the receiver is removed as thermal energy.

Specular

Having the qualities of a mirror. Angle of incidence equals angle of reflection.

Spillage (Radiation)

Radiation reflected from the collector subsystem, but which misses the absorber surface of the receiver.

Stand-Alone

A solar thermal central receiver power system that operates on solar energy only, with no on-site back-up power system.

Storage Capacity

The amount of net energy that can be delivered from a fully charged storage subsystem and be used as a source of energy to generate electricity. (Units: J or MW(th)-hr)

Storage-Coupled

The use of an energy storage system to permit operation of the end use system during periods when solar power from the receiver is inadequate (or not present) to satisfy the load.

Storage Utilization Factor

Fraction of the storage thermal capacity that can be delivered at rated conditions (i.e., at nearly constant thermodynamic quality).

Stow

A position or act of reaching a position of storage for heliostats or other movable collectors.

Thermal Energy Storage Subsystem

Any rechargeable unit capable of storing thermal energy for later use. Examples are sensible heat storage in nitrate salt, sodium, rocks, water, or oil.

Thermal Oil

A high temperature oil used as a thermal transfer and storage medium. For the DCS, this is a commerical oil produced by Monsanto (Santotherm 55).

Thermal Ratchetting

The growth in a dimension of a material due to thermal cycling. In the context of thermal energy storage subsystems using granular media, it refers to the settling of the media stored in the tank with rising tempearture (permitted by differential thermal expansion) followed by overstressing and stretching of the tank wall when the temperature drops.

Thermal Stratification Tank

(See *Thermocline Storage*)

Thermocline Storage

The storage of thermal energy where the hot and cold media are in the same container (tank) using the thermocline principle which relies on a lower density hot fluid floating atop a higher density cooler fluid of the same type or which relies on hot solid material being separated from cooler solid materials by a thermal gradient as in air/rock, air/ceramic brick applications. The *thermocline* is the zone or layer in which the vertical temperature profile changes rapidly.

Topping Cycle

The higher temperature cycle in any energy conversion system where two (or more) separate cycles are used in cascade fashion (exhaust of one feeds input to another). See *Bottoming Cycle.*

Trace Heating

See *Heat Tracing.*

Tracking System

The motors, gears, and actuators that are instructed by computer command to maintain a proper heliostat orientation with respect to the sun and receiver positions.

Utilization Factor (Collection Field)

The utilization factor of the collector field is the fraction of time when beam irradiance exceeds the threshold value that the collector field is operable.

Working Fluid

The fluid that performs work and that is utilized in the end-use system, e. g., the steam in a steam turbine-generating system. This may be the same as the receiver or heat transport fluid.

Solar Radiation

Air Mass, m

The path length of radiation through the atmosphere considering the vertical path at sea level as unity. Thus, at sea level, $m = 1$ when the sun is at zenith (directly overhead). Except for very large zenith angles ($m > 3$ where atmospheric refraction becomes significant) $m = \sec a$.

Albedo

The ratio of radiation reflected by a surface to that incident upon it.

Angle of Incidence

The angle between the central ray incident on a surface and the normal to the surface at the point of incidence.

Attenuation Loss

Loss of solar power by absorption and scattering due to atmospheric conditions. Losses are caused by scattering by air molecules, by selective absorption by certain molecules (e. g. water vapor), and by absorption and scattering by aerosols.

Beam Irradiance (Direct Beam Irradiation)

Unscattered focusable solar radiation (over density). See *Direct Beam Solar Radiation.* (Units: W m^{-2})

Beam Irradiation (Direct Beam Irradiation)

Incident solar energy from direct beam irradiation, defined for a specified period (*e. g.* daily, monthly). (Units: J m^{-2})

Circumsolar Radiation

Solar radiation scattered by the atmosphere into the area of the sky immediately adjacent to the sun. It produces the *solar aureole,* whose angular extent is directly related to the atmospheric turbidity, increasing with high turbidity. The amount and character of circumsular radiation vary widely with geographic locale, climate, season, time of day, and observing wavelength. The effect of circumsolar radiation on focusing collectors is of interest due to the overestimates of direct beam solar radiation that can result from use of pyrheliometer measurements and the dependence of the performance of a focusing collector on the details of the distribution of light near the solar disc.

Cloud Cover

That portion of the sky cover which is attributed to clouds, usually measured by a trained observer in tenths of the sky covered.

Cloudiness[13]

The amount of sky covered by clouds. Cloud amounts are typically measured in eights or tenths; the clouds do not have to be opaque. A rough estimate of cloudiness can be made from solar radiation measurements and vice versa.

Declination Angle

The angle between the sun's rays and the zenith direction (directly overhead) at noon on the earth's equator. It has the same numerical value as the latitude at which the sun is directly overhead at noon on a given day. By convention the solar declination angle is positive when the earth-sun vector points northward relative to the equatorial plane. The declination angle varies from $-23.45°$ on December 21 (the winter solstice) to $+23.45°$ on June 22 (the summer solstice). (All references are for the Northern hemisphere). If x is the solar declination angle, x can be determined within approximately 1° by

$$\sin x = 0.39795 \cos [0.98563 (N\text{-}173)]$$

where N is the day number of the year.

Design Insolation

The value of direct beam irradiance at which the plant is designed to produce its rated power. For the SSPS plants, this value is $920 \, W/m^2$ (equinox noon).

Diffuse Irradiance

The irradiance of the solar radiation scattered from the sky onto a receiving surface is called *diffuse radiation* or *sky radiation*. It is the downward scattered and reflected solar radiation coming from the entire hemisphere with the exception of the solid angle subtended by the sun's disc. The magnitude of diffuse sky radiation depends on solar elevation, the amount and type of aerosols present, and clouds. It is also influenced somewhat by the albedo of the earth's surface since radiation reflected from the ground is partially back-scattered from the atmosphere.

Diffuse Irradiation

The diffuse solar energy incident on a specified surface for a specified period of time. (Units: $J \, m^{-2}$)

Diffusograph

A pyranometer equipped with a special shading or occulting device to exclude direct beam solar radiation.

Direct (beam) Solar Radiation

The solar energy incident on a surface that comes from within the solid angle subtended by the solar disk, *i. e.,* that sunshine capable of casting a sharply defined shadow. The direct component of the insolation can be focused by an optical system. It is distinguished from the diffuse or multidirectional components of solar radiation. Cloud, fog, haze, smoke, dust, and molecular scattering increase the diffuse component.

Direct Normal Insolation

The direct beam radiation or insolation on a surface perpendicular to the sun's rays.

Duration of Sunshine[14]

The time (hours or minutes) during which the sunshine is intense enough to throw a shadow. Measurements using this imprecise criterion have been taken for over 140 years in Europe, and serve both as a general measure of regional cloudiness and as a rough measure of global irradiance. Models have been developed[15] that relate sunshine duration rasonably well with global irradiance. Such empirical models are useful for areas with few available solar radiation measurements, but are poor substitutes for actual measurements.

An empirical relationship[15]

$$G = G_o[a + b(S/S_o)]$$

where

G = Average global solar irradiance received for a specified location and period of time.

G_o = Horizontal extraterrestrial radiation for the same period and location.

S = Average daily hours of bright sunshine for the same period and location.

S_o = Maximum daily hours of bright sunshine for the same period and location.

a, b = Modified constants

provides a reasonable measure of global radiation from cloudiness or sunshine duration measurements. Extensive measurements in Austria[16] indicate that calculated global irradiance is within an error band of 34% for 99% of the cases.

Elevation Angle

The solar elevation angle in degrees is 90 – Zenith Angle

Extraterrestrial Solar Radiation

Solar radiation received at the upper limit of the earth's atmosphere. This is the solar radiation that would reach the ground in the absence of a planetary atmosphere.

Flux (Radiant)

The time rate of flow of radiant energy. (Units: W)

Flux Density

The radiant flux incident per unit area. (Units: $W\ m^{-2}$)

Global Irradiance or Global Solar Flux Density

The direct plus diffuse solar irradiance on a surface of any defined orientation and surrounding. It includes radiation reflected from the surroundings incident on the receiving surface. If the orientation of the receiving surface is not specified, *global* conventionally refers to a horizontal surface. (Units: $W\ m^{-2}$)

Global Irradiation

The global irradiance integrated over a specified time period. *Monthly Mean Daily Global Radiation* refers to the daily mean global irradiation on a horizontal surface exposed to a hemispherical sky, averaged over a specified month. (Units: $J\ m^{-2}$)

Global Radiation

The sum of direct and diffuse radiation from the sky. It is usually defined for a horizontal receiving surface exposed to a hemispherical sky (solid angle of 2 pi). (Units: W m^{-2}).

Insolation

Acronym for *Incoming Solar Radiation*

Intensity

The radiant flux leaving a source per unit solid angle (Units: W sr^{-1}). *Intensity* is also often used colloquially for *irradiance*.

Irradiance (or Flux Density)

The radiant flux incident per unit area of surface. (Units: W m^{-2})

Irradiation or Radiant Exposure

The radiant energy incident per unit area. The product of flux density (irradiance) and its duration. (Units: J m^{-2})

Minimum Operation Insolation

The insolation level below which a solar power plant cannot operate. For the DCS system, this level is approximately 300 W/m^2 (direct beam irradiance).

Pyroheliometer[17]

An instrument for measuring the intensity of direct beam solar radiation at normal incidence.

Pyranometer

An instrument for the measurement of the solar radiation received from the entire hemisphere.

Pyrradiometer

An instrument for the measurement of both solar and terrestrial radiation.

Radiance (Radiant Intensity)

The radiant flux leaving or arriving at a surface in a given direction per unit solid angle and per unit of surface area projected orthogonal to that direction. (Units: W m^{-2} sr^{-1})

Radiant Energy

Electromagnetic energy produced by a radiating source. (Units: J)

Radiant Exposure (Irradiation)

Radiant energy incident on a surface for a specified period of time. (Units: J m^{-2})

Radiant Flux Density

The electromagnetic power density. (Units: W m^{-2})

Radiant Flux or Radiant Power

The time rate of flow of radiant energy. (Units: W)

Radiation

The emission and propagation of energy through space (or through material medium) in the form of waves (or photons).

Solar Constant

Also called the *solar parameter* in recent terminology, this is the radiant flux density of solar radiation incident on a surface normal to the solar beam at the outer limit of the atmosphere, with the earth at its mean distance $(1.495 \times 10^{11}\,\mathrm{m})$ from the sun. The value accepted in 1980 by the IEA[18] is $1{,}373\,\mathrm{W\,m^{-2}}$.

Solar Energy

The energy transmitted from the sun in the form of electromagnetic radiation, normally perceived as sunshine, but encompassing a wider spectrum that seen by the human eye or transmitted by the atmosphere.

Solar Time

The time as reckoned by the apparent position of the sun. Solar noon is the instant at which the sun reaches its zenith. The diurnal cycle, reflecting the rotation of the earth about its axis, is the basis for the concept of *solar time*. A *solar day* is defined as the interval of time from the moment the sun crosses the local meridian to the next time it crosses that same meridian. (The local meridian at any point is the plane formed by projecting a north-south longitude line through that point out into space from the earth's center.) Because of the earth's forward movement in its orbit during this internval, the time required for one full rotation of the earth is less than a solar day by about 3.85 minutes.

The solar day varies in length through the year. This variance is due to the tilt of the earth's axis with respect to the plane of the ecliptic (the plane defined by the earth's orbit around the sun) and the ellipticity of the earth's orbit. *Solar noon,* the time when the sun crosses the local meridian, differs with longitude. Consequently, *standard time* (which is a uniform time) and *solar time* differ. This difference is called *the equation of time (EOT)*. It varies with date and longitude, and is given approximately by:

$$\mathrm{EOT} = 12 + 0.1236 \sin x - 0.0043 \cos x + 0.1538 \sin 2x + 0.0608 \cos 2x$$

where the angle x is a function of the day N of the year:

$$x = 360\,(N\text{-}1)/365.242 \text{ (degrees)}$$

The relationship between *local standard time* (LST) and *solar time* is given by:

$$\text{Solar Time} = \mathrm{LST} - \mathrm{EOT} - \mathrm{L}$$

where

$$\mathrm{L} = (1/15)\,[\text{local longitude} - \text{longitude of local time meridian}]$$

Sun Position

The azimuth and elevation angles for specifying the direction antiparallel to the central ray from the sun.

Sunshine Hours
(See *Duration of Sunshine*)

Zenith Angle
The angle subtended by the zenith and the line of sight to the sun.

Financial Terms

Avoided Cost
The cost not incurred by a utility to generate its own electricity as a result of the purchase of electricity from an independent producer. Over the short term, this amount is equivalent to the operating cost of the marginal company-owned generating unit or units. Over the long term, avoided cost equals the cost of constructing and operating new generating capacity.

Capacity Displacement
The amount of conventional generating capacity which may be omitted from a utility's planned requirements if a solar thermal electric plant is incorporated.

Capacity Payment
That portion of avoided cost reflecting the value of a plant's capacity rating to the utility system.

"Captive" Project
An energy project located at an industrial or commercial project and designed to provide energy only to that facility, displacing purchases by the facility from the electric utility.

Discount Rate
The annual rate used in present worth analyses to take into account inflation and the potential earning power of money while moving it forward or backward to a single point in time for comparison of value.

Economic Value
The marginal cost of energy and capacity displaced by a solar thermal electric project.

Energy Payment
That portion of avoided cost reflecting the amount of electrical energy not needed to be generated due to the operation of a solar thermal electric plant.

Fixed Charge Rate
The annual revenue requirements for return on capital, depreciation, and taxes, as a percentage of capital investment. Permits annualizing capital costs for comparison with annual operating costs.

Internal Rate of Return (IRR) Method
Acceptance criterion when the discount rate that equates the present value of the expected cash outflows with the present value of the cash inflows equals or exceeds a specified rate of return.

Levelization
The process by which a series of non-uniform future payments is converted into a uniform (level) series of payments whose present worth is equal to that of the original non-uniform series.

Levelized Fixed Charge Rate
The fixed charge rate that produces a constant level of payments over the life of the plant whose present worth is the same as the present worth of the actual cash flow.

Levelized Busbar Energy Cost
The constant annual revenue per unit of energy required over the life of a plant to compensate for its fixed and variable costs (mills per kWh).

Net Present Value (NPV) Method
Investment criterion for acceptance of an initial investment when the present value of the cash inflows equals or exceeds the present value of the cash outflows (including initial investment) at a specified rate of return.

Payback Period
Number of years required to recover an initial cash investment (no discounting).

Present Value (Present Worth)
The present value of a series of equal (annual) future payments discounted at a specified rate of return.

Risk Rate
Required rate of return on equity on positive cash flows commensurate with the perceived risk.

Safe Rate
The discount rate for which negative cash flows equivalent to the safe rate of return, such as U.S. Treasury bills. Discounting at the safe rate is equivalent to establishing a sinking fund for future obligations.

Standard Offer
A standard, predetermined and pre-approved utility contract (in the U.S.) for the purchase of electricity from an independent power producer.

Notes and References to Glossaries

1 P. B. Bos and J. M. Weingart (1983), *Bonneville Power Administration Comparative Electric Generation Study, Vol. II – Solar Thermal Central Receiver, Photovoltaic, and Wind Electric Technologies* Report 82-028-RE. Oakland, CA: Kaiser Engineers Power Corporation. (Available from NTIS).

2 P. B. Bos and J. M. Weingart (1983), *Impact of Tax Incentives on the Commercialization of Solar Thermal Electric Technologies*. Report SAND 83-8178. Livermore, CA: Sandia National Laboratories.

3 F. Kreith and J. F. Kreider (1978). *Principles of Solar Engineering*. New York: McGraw-Hill Book Company.

4 Sandia National Laboratories (1982). *Characteristics of Current Solar Central Receiver Projects*. International Workshop on the Design, Construction, and Operation of Solar Central Receiver Projects.

5 Sandia National Laboratories (1983). *Department of Energy Solar Central Receiver Annual Meeting*. SAND 83-8018.

6 International Energy Agency (1985). *Small Solar Power Systems (SSPS) Evaluation Report*.

7 DFVLR (1985). *Proceedings of the IEA-SSPS Experts Meeting on High Temperature Technology and Application*. June 18–21, 1985. SSPS Technical Report No. 1/85.

8 E. C. Boes (1979). *Fundamentals of Solar Radiation*. Sandia National Laboratories Report SAND 79-0490.

9 N. Robinson, ed. (1966). *Solar Radiation*. Amsterdam: Elsevier Publishing Company.

10 International Energy Agency (1980). *An Introduction to Meterological Measurements and Data Handling for Solar Energy Applications*. Report DOE/ER-0084. Washington, DC: U.S. Department of Energy, Office of Energy Research and Office of Basic Sciences.

11 International Energy Agency (1985). *Small Solar Power Systems Project Final Evaluation Report, Volume III: Meteorological and Environmental Conditions*. (Draft).

12 P. Bos, M. R. Eaton, and J. M. Weingart (1985), *Photovoltaic Systems Financing Guidebook*. Albuquerque, NM: Sandia National Laboratories.

13 World Meteorological Organization (1971). *Guide to Meteorological Instrument and Observing Practives*. No. 8, TP. 3.

14 M. R. Riches (1980). *Duration of Sunshine*. In *An Introduction to Meteorological Measurements and Data Handling for Solar Energy Applications*. International Energy Agency.

15 K. L. Coulson (1975). *Solar and Terrestrial Radiation: Methods and Measurements*. New York: Academic Press.

16 F. Neuwirth (1979). *Meteorological Research at Service of Solar Energy Utilization*. Results of Research and Development work in Austria, 1979, pp. 11–21. ASSA Report.

17 B. B. Williams (1980). *Meteorological Variables Related to Solar Energy*. In *An Introduction to Meteorological Measurements and Data Handling for Solar Energy Applications*. International Energy Agency.

18 International Energy Agency (October, 1980). *An Introduction to Meteorological Measurements and Data Handling for Solar Energy Applications*. Report DOE/ER-0084. U.S. Department of Energy.

References and Literature

I: INTERNATIONAL ENERGY AGENCY, SMALL SOLAR POWER
            SYSTEMS PROJECT: "The IEA/SSPS Solar Thermal Power Plants",
            Vol. I: Central Receiver System. (numerous authors, sections 1 through 9
            and appendices)
II: INTERNATIONAL ENERGY AGENCY, SMALL SOLAR POWER
            SYSTEMS PROJECT: "The IEA/SSPS Solar Thermal Power Plants",
            Vol. II; Distributed Collector System (numerous authors, sections 1 through
            8)
III   INTERNATIONAL ENERGY AGENCY, SMALL SOLAR POWER
            SYSTEMS PROJECT: "The IEA/SSPS Solar Thermal Power Plants", Vol-
            . III; Site Specifics (numerous authors, sections 1 through 5)
IV: INTERNATIONAL ENERGY AGENCY, SMALL SOLAR POWER
            SYSTEMS PROJECT: "The IEA/SSPS Solar Thermal Power Plants",
            Vol. IV: Book of Summaries (numerous authors)
CA: Contributed by author
EFN: Energiforskningsnamden (Swedish) Projektresultat Efn/LET by Jonas Sand-
            gren and Mats Andersson
SR 1 *:   IEA Small Solar Power Systems Project SR 1, "DCS Construction Report"
            by A. Kalt, M. Loosme and H. Dehne
SR 2 *:   IEA Small Solar Power Systems Project SR 2, "CRS Construction Report"
            by M. Becker, H. Ellgering and D. Stahl
SR 3 *:   IEA Small Solar Power Systems Project SR 3, "DCS First Year of Opera-
            tion" by A. Kalt
SR 4 *:   IEA Small Solar Power Systems Project SR 4, "SSPS-CRS First Period of
            Operation" by W. Bucher
SR 5 *:   IEA Small Solar Power Systems Project SR 5, "ASR Construction Experi-
            ence Report" by John Hansen
SR 6 *:   IEA Small Solar Power Systems Project SR 6, "DCS Supplement / Construc-
            tion Experience Report" by John Hansen
SR 7 *:   IEA Small Solar Power Systems Project SR 7, "SSPS Results of Tests and
            Operations /1981 – 1984" by Wilfried Grasse
TR 1/81 *: SSPS TECHNICAL REPORT No. 1/81
            Tabernas Meteo Data Analysis Based on Evaluated Data Prepared by the
            SSPS-O.A., by Belgonucleaire, Belgium.
TR 3/82 *: SSPS TECHNICAL REPORT No. 3/82
            Effect of Sunshape on Flux Distribution and Intercept Factor of the Solar
            Tower Power Plant at Almeria by G. Lemperle, DFVLR, Germany.
TR 3/83 *: SSPS TECHNICAL REPORT No. 3/83
            The Advanced Sodium Receiver (ASR) – Topic Reports – by Agip Nuc-
            leaire, Milano and Franco Tosi, Legnano, Italy.
TR 6/83 *: SSPS TECHNICAL REPORT No. 6/83
            First Year Average Performance of the SSPS-DCS Plant by Thierry J. van
            Steenberghe, IEA SSPS ITET.
TR 1/84 *: SSPS TECHNICAL REPORT No. 1/84
            Executive Summary IEA/SSPS – CRS Workshop (April 1983) by Clifford S.
            Selvage (SSPS-ITET), U.S.A.

1:       Personal communications from W. Grasse and C. S. Selvage, ITET
2:       "Solar Power Tower Design Guide: Solar Thermal Central Receiver Power Systems, a Source of Electricity and/or Process Heat" by K. W. Battleson, Sandia National Laboratories, Livermore California 94550, USA.
3:       "A Model for the Economic Assessment of Solar Power Plants" by W. Bucher, M. Geyer, H. Klaiss, J. P. Thornton. Chairman: G. Faninger, ASSA. Vienna, Austria.
4:       "Solarthermisches Elektrizitätswerk SOTEL" ISBN-3-85677-003-8 EIR, 1982 (numerous authors)
        CH 5303 Wurenlingen, Switzerland.
5:       "METAROZ, Studie über die Möglichkeiten eines Solarthermischen Kraftwerkes im VAL MAROZ", Study on the feasibility of a solar thermal power plant in the Maroz valley (Switzerland). (numerous authors)
        CH 5303 Wuerenlingen, Switzerland.
6:       "Experiences with Solar Power" by P. Kesselring Symposium "Solar Energy 85" in Igels (Austria), 7. 31 – 8. 9. 85 ESA SP – 240 Nov. 1985.
7:       "Thermo-Mechanical Solar Power Plants", Proceedings of the Second International Workshop on the Design, Construction and Operation of Solar Central Receiving Projects, Varese, Italy, 4–8 June 1984. Published for the Commission of European Communities by the D. Reidel Publishing Company, Dordrecht, Boston, Lancaster.
8:       "Plataforma Solar de Almeria. Projecto CESA-1" by J. Avellaner, F. Sanchez Sudon and C. Ortiz. Ministerio de Industria, JEN-Instituto de Energias Renovables, Madrid, Spain.
9:       "Shenandoah Solar Total Energy project' by E. J. Ney, manager solar operations, Georgia Power Co. 7 Solar circle, Shenandoah Georgia 30265 USA and D. M. Moore, Georgia Institute of Technology, Engineering experimant station, Atlanta Georgia 30332 USA.
10:      Helga Steeg, Director of the IEA, statement given on occasion of a lecture given in Bern, Switzerland, upon being invited by the "Swiss Energy Forum", a private asociation of members of the Swiss parliament, scientists and industrialists on October 3. 1985.
11:      "Solar Energy Utilization – a Technical, Economical and Political Task for an Industrialized Country in Middle Europe" by C. J. Winter, J. Nitsch, H. Klaisz and C. Voigt, DFVLR 7000 Stuttgart 80, German Federal Republic.
12:      "Impact of Tax Incentives on the Commercialization of Solar Thermal Electric Technologies", by P. B. Bos and J. M. Weingart, Polydine Associates, Menlo Park, California. Prepared by Sandia Laboratories, Livermore California 94550. USA.
+:   References indicated by roman numerals are obtainable in book-shops (Springer-Verlag Berlin Heidelberg New York Tokyo 1986)
        ISBN-No's 3-540-16145-7; 0-387-16145-7 (4 Volumes)
                3-540-16146-5; 0-387-16146-5 (Vol. 1)
                3-540-16147-3; 0-387-16147-3 (Vol. 2)
                3-540-16148-1; 0-387-16148-1 (Vol. 3)
                3-540-16149-x; 0-387-16149-x (Vol. 4)
*:   References indicated by capital letters may be obtained directly from the IEA SSPS project office at the following adress:
        DFVLR
        IEA SSPS Project
        Linder Höhe
        D-5000 Köln 90
        (Federal Republic of Germany)
        All numbered references may be obtained from the sources cited.

Executive Committee (EC) Members

The SSPS Project has been conducted in its first phase by 10 contracting parties and in its second phase by 9 parties (Great Britain did not continue in the project). Each party was represented by an EC-member and an alternate. These representatives may be contacted for further information concerning the contributions of their countries. The Project was managed by the German "Deutsche Forschungs- und Versuchsanstalt für Luft- und Raumfahrt e. V. (DFVLR)" as Operating Agent, efficiently supported by the Host Country, Spain.

The members are listed per country, the present member in italics former members appended in alphabetical order.

(A) Austria *Prof. Dr. G. Faninger* ASSA
(B) Belgium *F. Altdorfer* SPPS
J. C. Delcroix SPPS
T. Vijverman SPPS
(CH) Switzerland *Dr. P. Kesselring* EIR
(D) Federal Republic of Germany *Dr. W. Rasch* DFVLR
Prof. Dr. C.-J. Winter DFVLR
(E. C. Chairman from Oct. 1985 to date)
(E) Spain *L. Crespo Rodriguez* CIEMAT-IER
Dr. A. Munoz Torralbo CEE
Dr. C. Sanchez Lopez JEN-IER
J. Temboury CEE
(GB) Great Britain (Great Britain withdrew from the Project in 1978)
J. K. Duxbury
G. L. Shires
(GR) Greece (Greece withdrew from the Project in March 1986)
Dr. E. N. Carabateas NEC
(I) Italy *Prof. F. Reale* CNR-PFE
(S) Sweden *Dr. B. Finnström* ERC
I. Andersson NE
L. Brandels NE
K. B. Heden NE
L. Rey NE
(E. C. Chairman from 1977 to 1983)
(USA) United States *C. Carwile* DOE
G. W. Braun DOE
L. Gutierrez SANDIA
(E. C. Chairman from 1983 to 1985)

Name and Subject Index

128   Subject Index

P. Kesselring, C. S. Selvage (Eds.)

The IEA/SSPS Solar Thermal Power Plants

Facts and Figures/Final Report of the International
Test and Evaluation Team (ITET)

Volume 1: **Central Receiver System (CRS)**
Volume 2: **Distributed Collector System (DCS)**
Volume 3: **Site Specifics**
Volume 4: **Book of Summaries**

1986. XXX, 1639 pages. Volumes 1–4 (as a set).
Soft cover DM 350,–. ISBN 3-540-16145-7

This publication is the first, complete documentation
of the experiences, had with the design, construction,
and operation of two dissimilar (Farm type, Tower
type) solar thermal power plants. The project was
carried out under the auspieces of the International
Energy Agency, scientists, and engineers from 9 participating countries have contributed to the compilation
of these four volumes. By presenting all the papers
from the final evaluation workshop of the project, this
work is more than just a proceedings. Papers are divided into sections with commentaries and summaries
by the editors to provide quick reference to the 1500
papers of content.
The publication consists of three volumes, dealing
with the Central Receiver System (CRS), the
Distributed Collector System (DCS), the Site Specifics, and a fourth one, the Book of Summaries. It will
be especially helpful to scientists and engineers en-
gaged in research, system analysis, design, and
construction in the field of solar thermal energy
conversion and application.

Springer-Verlag
Berlin Heidelberg New York
London Paris Tokyo

Springer

Each volume also available separately

Volume 1

Central Receiver System (CRS)

1986. IX, 776 pages. Soft cover DM 160,-.
ISBN 3-540-16146-5

Contents: Introduction. – Central Receiver System
Description. – Historical Assessment of CRS Plant
Performance and Operational Experience. – Heliostat
Field Performance. – Receiver Behavior Comparison. –
Thermal Losses/Thermal Inertia. – Systems Aspects/
Control. – Potential for Improvements. – Appendices.

Volume 2

Distributed Collector System (DCS)

1986. VIII, 510 pages. Soft cover DM 110,–
ISBN 3-540-16147-3

Contents: Introduction. – Distributed Collector System.
– Historical Assessment of the SSPS-DCS Plant
Performance – Introduction. – Survey of Plant Losses –
Introduction. – Possibility of Automatic Control – Intro-
duction. – Reliability – Availability – Maintenance –
Introduction. Potential for Improvements. – Appendi-
ces.

Volume 3

Site Specifics

1986. VII, 192 pages. Soft cover DM 60,–
ISBN 3-540-16148-1

Contents: Introduction. – Site Description. – Meteorological Conditions. – Environmental Conditions/
Reflectivity. – Soiling.

Volume 4

Book of Summaries

Springer-Verlag
Berlin Heidelberg New York
London Paris Tokyo

1986. VI, 161 pages. Soft cover DM 60,–
ISBN 3-540-16149-X

Contents: Summaries of all Chapters of Volumes 1–3.